CLARENDON LIBRARY OF LOGIC AND PHILOSOPHY

General Editor: L. Jonathan Cohen

THE SCIENTIFIC IMAGE

The *Clarendon Library of Logic and Philosophy* brings together books, by new as well as by established authors, that combine originality of theme with rigour of statement. Its aim is to encourage new research of a professional standard into problems that are of current or perennial interest.

Also published in this series

THE
SCIENTIFIC
IMAGE

BAS C. VAN FRAASSEN

CLARENDON PRESS · OXFORD

Oxford University Press, Walton Street, Oxford OX2 6DP

Oxford New York
Athens Auckland Bangkok Bombay
Calcutta Cape Town Dar es Salaam Delhi
Florence Hong Kong Istanbul Karachi
Kuala Lumpur Madras Madrid Melbourne
Mexico City Nairobi Paris Singapore
Taipei Tokyo Toronto
and associated companies in
Berlin Ibadan

Oxford is a trade mark of Oxford University Press

Published in the United States by
Oxford University Press Inc., New York

British Library Cataloguing in Publication Data
Van Fraassen, Bas C.
The scientific image.—(Clarendon library of logic and philosophy)
1. Science—Philosophy
I. Title
501   Q175   79-42793
ISBN 0-19-824427-4 Pbk.

9 10 8

Printed in Great Britain
on acid-free paper by
Biddles Ltd
Guildford and King's Lynn

To the friendly people of Tigh-na-Coille

Preface

THE aim of this book is to develop a constructive alternative to scientific realism, a position which has lately been much discussed and advocated in philosophy of science. To this end, I shall present three theories, which need each other for mutual support. The first concerns the relation of a theory to the world, and especially what may be called its empirical import. The second is a theory of scientific explanation, in which the explanatory power of a theory is held to be a feature which does indeed go beyond its empirical import, but which is radically context-dependent. And the third is an explication of probability as it occurs within physical theory (as opposed to: in the evaluation of its evidential support). The first two chapters form a brief and relatively popular introduction to the debates concerning scientific realism, and will thereby explain the organization and strategy of the remainder. I have kept the exposition non-technical throughout, referring for technical details to journal articles where they seem to me more rightfully to belong.

My debts are numerous; many of them will be clear from the notes. I would like to add here a few personal acknowledgements. My greatest debt in philosophy of science has always been to Adolf Grünbaum: this debt was renewed when I attended his lecture on Dirac's electrodynamics at Santa Margharita in 1976, a paradigm of philosophical exposition of science which I can scarcely hope to emulate. To Glymour, Hooker, Putnam, Salmon, Smart, and Sellars I owe the debt of the challenge of their philosophical positions and their willingness to discuss them with me, both on public occasions and in personal correspondence. The title of this book is a phrase of Wilfrid Sellars's, who contrasts the scientific image of the world with the manifest image, the way the world appears in human observation. While I would deny the suggestion of dichotomy, the phrase seemed apt. Toraldo di Francia gave me the opportunity to take part in the Fermi Institute Summer School on Foundations of Physics in Varenna, where I learned a great deal, not least from his and Dalla Chiara's lectures on their theory of the structure of physics. An older debt recalled in writing various parts of this book is to Henry Margenau, from whom I learned much about prob-

abilities and states in quantum mechanics. Many friends and colleagues helped at various stages during the writing of this book by
reacting, at once sympathetically and ruthlessly, to my arguments,
ideas, and didactic stories; Paul Benacerraf, Nancy Cartwright,
Ronald de Sousa, Hartry Field, Yvon Gauthier, Ronald Giere,
Karel Lambert, Edwin Levy, Margot Livesey, Hugh Mellor, Ben
Rogers, Richmond Thomason, and Roger Woolhouse, to mention
only a few. The main theses of this book were presented in lectures
at a number of occasions, the last before press being three lectures
at Princeton University in May 1979. Finally, the aid of the Canada
Council in support of the research projects during which this book
took shape was invaluable, especially in facilitating the contacts with
other scholars necessary for that research.

B. C. v. F.

July, 1979

Contents

1
Introduction

[I]t is easy to indulge the commonplace metaphysical instinct. But a taste for metaphysics may be one of those things which we must renounce, if we mean to mould our lives to artistic perfection. Philosophy serves culture, not by the fancied gift of absolute or transcendental knowledge, but by suggesting questions ...

Walter Pater, *The Renaissance*

THE opposition between empiricism and realism is old, and can be introduced by illustrations from many episodes in the history of philosophy. The most graphic of these is perhaps provided by the sense of philosophical superiority the participants in the early development of modern science felt toward the Aristotelian tradition. In that tradition, the realists held that regularities in the natural phenomena must have a reason (cause, explanation), and they sought this reason in the causal properties, constituting what they called the substantial forms or natures, of the substances involved in natural processes. The nominalists, who denied the reality of these properties, were in the position of having to reject such requests for explanation.[1]

The philosophers engaged in developing the philosophical foundations of modern science had apparently escaped this dilemma. Without postulating such causal properties, forms, or 'occult qualities', they could still explain the regularities that are observed in nature. Thus Robert Boyle writes,

That which I chiefly aim at, is to make it probable to you by experiments, that almost all sorts of qualities, most of which have been by the schools either left unexplicated, or generally referred to I know not what incomprehensible substantial forms, may be produced mechanically; I mean by such corporeal agents as do not appear either to work otherwise than by virtue of the motion, size, figure, and contrivance of their own parts (which attributes I call the mechanical affections of matter).[2]

To give an account of such phenomena as heat or chemical reactions in terms only of mechanical attributes, they realized quite well, required at least an atomic theory of matter. But I suppose it is clear that they will face that same dilemma again for the regularities they postulate in the behaviour of the atomic parts. No mechanical explanations are possible there, since the atoms have no further parts. So either they must attribute specific powers, qualities, and causal properties to those atoms to explain why they act and react in the way they actually do; or else they must, like the nominalists before them, reject the request for explanation.

In addition, they have gained a problem. Part of the motivation for the nominalist rejection of the Aristotelian realists' world of powers, properties, dispositions (made famous by Molière's *virtus dormitiva*) was epistemological. The observation of the phenomena did not point unambiguously to the supposed causal connections behind them. This problem exists similarly for the atomic hypotheses: the phenomena do not decide their truth or falsity, though they are perhaps better explained by one hypothesis than by another. Subsequent scientists intent on clarifying the philosophical basis of their discipline found it ever more difficult to reconcile their professed empiricism and antipathy to metaphysics with an unqualified belief in hypotheses that describe a supposed world behind the phenomena.

This led in the nineteenth century to the phenomenalism of Ernst Mach, the conventionalism of Henri Poincaré, and the fictionalism of Pierre Duhem. In the twentieth, the logical empiricism of Hans Reichenbach and logical positivism of Rudolph Carnap were further developments in this radical turn to empiricism.

Today, however, no one can adhere to any of these philosophical positions to any large extent. Logical positivism, especially, even if one is quite charitable about what counts as a development rather than a change of position, had a rather spectacular crash. So let us forget these labels which never do more than impose a momentary order on the shifting sands of philosophical fortune, and let us see what problems are faced by an *aspirant* empiricist today. What sort of philosophical account is possible of the aim and structure of science?

Studies in philosophy of science divide roughly into two sorts. The first, which may be called foundational, concerns the content and structure of theories. The other sort of study deals with the relations of a theory on the one hand, to the world and to the theory-user on the other.

There are deep-going philosophical disagreements about the

general structure of scientific theories, and the general characterization of their content. A current view, not altogether uncontroversial
but still generally accepted, is that theories account for the phenomena (which means, the observable processes and structures) by
postulating other processes and structures not directly accessible to
observation; and that a system of any sort is described by a theory
in terms of its possible states. This is a view about the structure of
theories shared by many philosophers who nevertheless disagree on
the issues concerning a theory's relation to the world and to its users.
Opponents of that view will at least say, I think, that this account
of what science is like is true 'on the face of it', or correct as a first
approximation.

One relation a theory may have to the world is that of being true,
of giving a true account of the facts. It may at first seem trivial to
assert that science aims to find true theories. But coupled with the
preceding view of what theories are like, the triviality disappears.
Together they imply that science aims to find a true description of
unobservable processes that explain the observable ones, and also of
what are possible states of affairs, not just of what is actual.
Empiricism has always been a main philosophical guide in the study
of nature. But empiricism requires theories only to give a true
account *of what is observable*, counting further postulated structure
as a means to that end. In addition empiricists have always eschewed
the reification of possibility (or its dual, necessity). Possibility and
necessity they relegate to relations among ideas, or among words, as
devices to facilitate the description of what is actual. So from an
empiricist point of view, to serve the aims of science, the postulates
need not be true, except in what they say about what is actual and
empirically attestable.

When this empiricist point of view was represented by logical
positivism, it had added to it a theory of meaning and language, and
generally a linguistic orientation. Today that form of empiricism is
opposed by scientific realism, which rejects not only the views on
meaning of the positivists, but also those empiricist tenets which I
outlined in the preceding paragraph. My own view is that empiricism
is correct, but could not live in the linguistic form the positivists
gave it. They were right to think in some cases that various philosophical perplexities, misconceived as problems in ontology and
epistemology, were really at bottom problems about language. This
opinion is correct especially, I think, about problems concerning

possibility and necessity. The language of science, being a proper part of natural language, is clearly part of the subject of general philosophy of logic and language. But this only means that *certain* problems can be set aside when we are doing philosophy of science, and emphatically does *not* mean that philosophical concepts must be one and all linguistically explicated. The logical positivists, and their heirs, went much too far in this attempt to turn philosophical problems into problems about language. In some cases their linguistic orientation had disastrous effects in philosophy of science. Scientific realism, however, pursues the antithetical error of reifying whatever cannot be defined away.

Correlative to discussions of the relation between a theory and the world, is the question what it is to accept a scientific theory. This question has an epistemic dimension (how much belief is involved in theory acceptance?) and also a pragmatic one (what else is involved besides belief?). On the view I shall develop, the belief involved in accepting a scientific theory is only that it 'saves the phenomena', that is, correctly describes what is observable. But acceptance is not merely belief. We never have the option of accepting an all-encompassing theory, complete in every detail. So to accept one theory rather than another one involves also a commitment to a research programme, to continuing the dialogue with nature in the framework of one conceptual scheme rather than another. Even if two theories are empirically equivalent, and acceptance of a theory involves as belief only that it is empirically adequate, it may still make a great difference which one is accepted. The difference is pragmatic, and I shall argue that pragmatic virtues do not give us any reason over and above the evidence of the empirical data, for thinking that a theory is true.

So I shall argue for an empiricist position, and against scientific realism. In some ways, philosophy is a subject of fashions—not more so than other intellectual disciplines, I suppose, but at least to the extent that almost any philosopher will begin by explaining that he opposes the 'dominant' or 'received' view, and present his own as revolutionary. It would be quite suspicious therefore if I were to say at this point that scientific realism has become dominant in philosophy of science. Others have certainly characterized it as the emerging victor: Isaac Levi recently wrote, 'My own view is that the coffin of empiricism is already sealed tight.'[3] And Arthur Fine, in a reply to Richard Healey:

The objections that he raises to a realist understanding of [quantum mechanics] are ... supposed to move my philosophical colleagues to the same anti-realist convictions that Mr. Healey thinks are held by many physicists. I am not sure how many physicists do hold such anti-realist convictions these days ... I suspect ... that most physicists who do shy away from realism are influenced more by the tradition in which they are schooled than they are by these rather recent and sophisticated arguments. That tradition is the deeply positivist legacy of Bohr and Heisenberg ... I am not much worried that my philosophical colleagues will be seduced by positivist considerations coupled with insubstantial reasons, for we are differently schooled.[4]

There is therefore at least already considerable sentiment on the side of realists that they have replaced the ametaphysical empiricism of the positivists. The empiricist position I mean to advocate will be strongly dissociated from both. (See Chapter 2 §1.2 and Chapter 3 §6 for some remarks on positivism.)

In part my argument will be destructive, countering the arguments brought forward by scientific realists against the empiricist point of view. I shall give a momentary name, 'constructive empiricism', to the specific philosophical position I shall advocate. The main part of that advocacy will be the development of a constructive alternative to scientific realism, on the main issues that divide us: the relation of theory to world, the analysis of scientific explanation, and the meaning of probability statements when they are part of a physical theory. I use the adjective 'constructive' to indicate my view that scientific activity is one of construction rather than discovery: construction of models that must be adequate to the phenomena, and not discovery of truth concerning the unobservable. The baptism of this philosophical position as a specific 'ism' is not meant to imply the desire for a school of thought; only to reflect that scientific realists have appropriated a most persuasive name for themselves (aren't we all scientific, and realists, nowadays?), and that there is after all something in a name.

2

Arguments Concerning Scientific Realism

> The rigour of science requires that we distinguish well the
> undraped figure of nature itself from the gay-coloured
> vesture with which we clothe it at our pleasure.
> Heinrich Hertz, quoted by Ludwig Boltzmann,
> letter to *Nature*, 28 February 1895

IN our century, the first dominant philosophy of science was developed as part of logical positivism. Even today, such an expression as 'the received view of theories' refers to the views developed by the logical positivists, although their heyday preceded the Second World War.

In this chapter I shall examine, and criticize, the main arguments that have been offered for scientific realism. These arguments occurred frequently as part of a critique of logical positivism. But it is surely fair to discuss them in isolation, for even if scientific realism is most easily understood as a reaction against positivism, it should be able to stand alone. The alternative view which I advocate —for lack of a traditional name I shall call it *constructive empiricism* —is equally at odds with positivist doctrine.

§1. *Scientific Realism and Constructive Empiricism*

In philosophy of science, the term 'scientific realism' denotes a precise position on the question of how a scientific theory is to be understood, and what scientific activity really is. I shall attempt to define this position, and to canvass its possible alternatives. Then I shall indicate, roughly and briefly, the specific alternative which I shall advocate and develop in later chapters.

§1.1 *Statement of Scientific Realism*

What exactly is scientific realism? A naïve statement of the position would be this: the picture which science gives us of the world is a true one, faithful in its details, and the entities postulated in science really exist: the advances of science are discoveries, not inventions. That statement is too naïve; it attributes to the scientific realist the belief that today's theories are correct. It would mean that the philosophical position of an earlier scientific realist such as C. S. Peirce had been refuted by empirical findings. I do not suppose that scientific realists wish to be committed, as such, even to the claim that science will arrive in due time at theories true in all respects— for the growth of science might be an endless self-correction; or worse, Armageddon might occur too soon.

But the naïve statement has the right flavour. It answers two main questions: it characterizes a scientific theory as a story about what there really is, and scientific activity as an enterprise of discovery, as opposed to invention. The two questions of what a scientific theory is, and what a scientific theory does, must be answered by any philosophy of science. The task we have at this point is to find a statement of scientific realism that shares these features with the naïve statement, but does not saddle the realists with unacceptably strong consequences. It is especially important to make the statement as weak as possible if we wish to argue against it, so as not to charge at windmills.

As clues I shall cite some passages most of which will also be examined below in the contexts of the authors' arguments. A statement of Wilfrid Sellars is this:

to have good reason for holding a theory is *ipso facto* to have good reason for holding that the entities postulated by the theory exist.[7]

This addresses a question of epistemology, but also throws some indirect light on what it is, in Sellars's opinion, to hold a theory. Brian Ellis, who calls himself a scientific entity realist rather than a scientific realist, appears to agree with that statement of Sellars, but gives the following formulation of a stronger view:

I understand scientific realism to be the view that the theoretical statements of science are, or purport to be, true generalized descriptions of reality.[1]

This formulation has two advantages: It focuses on the understanding of the theories without reference to reasons for belief, and it avoids the suggestion that to be a realist you must believe current scientific theories to be true. But it gains the latter advantage by use of the word 'purport', which may generate its own puzzles.

Hilary Putnam, in a passage which I shall cite again in Section 7, gives a formulation which he says he learned from Michael Dummett:

A realist (with respect to a given theory or discourse) holds that (1) the sentences of that theory are true or false; and (2) that what makes them true or false is something external—that is to say, it is not (in general) our sense data, actual or potential, or the structure of our minds, or our language, etc.[29]

He follows this soon afterwards with a further formulation which he credits to Richard Boyd:

That terms in mature scientific theories typically refer (this formulation is due to Richard Boyd), that the theories accepted in a mature science are typically approximately true, that the same term can refer to the same thing even when it occurs in different theories—these statements are viewed by the scientific realist ... as part of any adequate scientific description of science and its relations to its objects.[33]

None of these were intended as definitions. But they show I think that truth must play an important role in the formulation of the basic realist position. They also show that the formulation must incorporate an answer to the question what it is to *accept* or *hold* a theory. I shall now propose such a formulation, which seems to me to make sense of the above remarks, and also renders intelligible the reasoning by realists which I shall examine below—without burdening them with more than the minimum required for this.

Science aims to give us, in its theories, a literally true story of what the world is like; and acceptance of a scientific theory involves the belief that it is true. This is the correct statement of scientific realism.

Let me defend this formulation by showing that it is quite minimal, and can be agreed to by anyone who considers himself a scientific realist. The naïve statement said that science tells a true story; the correct statement says only that it is the aim of science to do so. The aim of science is of course not to be identified with individual scientists' motives. The aim of the game of chess is to checkmate your opponent; but the motive for playing may be fame, gold, and glory. What the aim is determines what counts as success in the enterprise as such; and this aim may be pursued for any number of reasons. Also, in calling something *the* aim, I do not deny that there are other subsidiary aims which may or may not be means to that end: everyone will readily agree that simplicity, informativeness, predictive power, explanation are (also) virtues. Perhaps my formulation can even be accepted by any philosopher who considers the most important aim of science to be something which only *requires* the finding of true theories—given that I wish to give the weakest formulation of the doctrine that is generally acceptable.

I have added 'literally' to rule out as realist such positions as imply that science is true if 'properly understood' but literally false or meaningless. For that would be consistent with conventionalism, logical positivism, and instrumentalism. I will say more about this below; and also in Section 7 where I shall consider Dummett's views further.

The second part of the statement touches on epistemology. But it only equates acceptance of a theory with belief in its truth.[2] It does not imply that anyone is ever rationally warranted in forming such a belief. We have to make room for the epistemological position, today the subject of considerable debate, that a rational person never assigns personal probability 1 to any proposition except a tautology. It would, I think, be rare for a scientific realist to take this stand in epistemology, but it is certainly possible.[3]

To understand qualified acceptance we must first understand acceptance *tout court*. If acceptance of a theory involves the belief that it is true, then tentative acceptance involves the tentative adoption of the belief that it is true. If belief comes in degrees, so does acceptance, and we may then speak of a degree of acceptance involving a certain degree of belief that the theory is true. This must of course be distinguished from belief that the theory is approximately true, which seems to mean belief that some member of a class centring on the mentioned theory is (exactly) true. In this way the proposed formulation of realism can be used regardless of one's epistemological persuasion.

§1.2 *Alternatives to Realism*

Scientific realism is the position that scientific theory construction aims to give us a literally true story of what the world is like, and that acceptance of a scientific theory involves the belief that it is true. Accordingly, anti-realism is a position according to which the aim of science can well be served without giving such a literally true story, and acceptance of a theory may properly involve something less (or other) than belief that it is true.

What does a scientist do then, according to these different positions? According to the realist, when someone proposes a theory, he

is asserting it to be true. But according to the anti-realist, the pro-
poser does not assert the theory to be true; *he displays it*, and
claims certain virtues for it. These virtues may fall short of truth:
empirical adequacy, perhaps; comprehensiveness, acceptability
for various purposes. This will have to be spelt out, for the details
here are not determined by the denial of realism. For now we must
concentrate on the key notions that allow the generic division.

The idea of a literally true account has two aspects: the language
is to be literally construed; and so construed, the account is true.
This divides the anti-realists into two sorts. The first sort holds that
science is or aims to be true, properly (but not literally) construed.
The second holds that the language of science should be literally
construed, but its theories need not be true to be good. The anti-
realism I shall advocate belongs to the second sort.

It is not so easy to say what is meant by a literal construal. The
idea comes perhaps from theology, where fundamentalists construe
the Bible literally, and liberals have a variety of allegorical,
metaphorical, and analogical interpretations, which 'demythologize'.
The problem of explicating 'literal construal' belongs to the philosophy of language. In Section 7 below, where I briefly examine some
of Michael Dummett's views, I shall emphasize that 'literal' does
not mean 'truth-valued'. The term 'literal' is well enough understood
for general philosophical use, but if we try to explicate it we find
ourselves in the midst of the problem of giving an adequate account
of natural language. It would be bad tactics to link an inquiry into
science to a commitment to some solution to that problem. The
following remarks, and those in Section 7, should fix the usage of
'literal' sufficiently for present purposes.

The decision to rule out all but literal construals of the language
of science, rules out those forms of anti-realism known as *positivism*
and *instrumentalism*. First, on a literal construal, the apparent
statements of science really are statements, *capable of* being true or
false. Secondly, although a literal construal can elaborate, it cannot
change logical relationships. (It is possible to elaborate, for instance,
by identifying what the terms designate. The 'reduction' of the
language of phenomenological thermodynamics to that of statistical
mechanics is like that: bodies of gas are identified as aggregates of
molecules, temperature as mean kinetic energy, and so on.) On the
positivists' interpretation of science, theoretical terms have meaning
only through their connection with the observable. Hence they hold

that two theories may in fact *say the same thing* although in form they contradict each other. (Perhaps the one says that all matter consists of atoms, while the other postulates instead a universal continuous medium; they will say the same thing nevertheless if they agree in their observable consequences, according to the positivists.) But two theories which contradict each other in such a way can 'really' be saying the same thing only if they are not literally construed. Most specifically, if a theory says that something exists, then a literal construal may elaborate on what that something is, but will not remove the implication of existence.

There have been many critiques of positivist interpretations of science, and there is no need to repeat them. I shall add some specific criticisms of the positivist approach in the next chapter.

§1.3 *Constructive Empiricism*

To insist on a literal construal of the language of science is to rule out the construal of a theory as a metaphor or simile, or as intelligible only after it is 'demythologized' or subjected to some other sort of 'translation' that does not preserve logical form. If the theory's statements include 'There are electrons', then the theory says that there are electrons. If in addition they include 'Electrons are not planets', then the theory says, in part, that there are entities other than planets.

But this does not settle very much. It is often not at all obvious whether a theoretical term refers to a concrete entity or a mathematical entity. Perhaps one tenable interpretation of classical physics is that there are no concrete entities which are forces—that 'there are forces such that . . .' can always be understood as a mathematical statement asserting the existence of certain functions. That is debatable.

Not every philosophical position concerning science which insists on a literal construal of the language of science is a realist position. For this insistence relates not at all to our epistemic attitudes toward theories, nor to the aim we pursue in constructing theories, but only to the correct understanding of *what a theory says*. (The fundamentalist theist, the agnostic, and the atheist presumably agree with each other (though not with liberal theologians) in their understanding of the statement that God, or gods, or angels exist.) After deciding that the language of science must be literally understood, we can still say that there is no need to believe good theories to be true, nor to believe *ipso facto* that the entities they postulate are real.

Science aims to give us theories which are empirically adequate; and acceptance of a theory involves as belief only that it is empirically adequate. This is the statement of the anti-realist position I advocate; I shall call it *constructive empiricism.*

This formulation is subject to the same qualifying remarks as that of scientific realism in Section 1.1 above. In addition it requires an explication of 'empirically adequate'. For now, I shall leave that with the preliminary explication that a theory is empirically adequate exactly if what it says about the observable things and events in this world, is true—exactly if it 'saves the phenomena'. A little more precisely: such a theory has at least one model that all the actual phenomena fit inside. I must emphasize that this refers to *all* the phenomena; these are not exhausted by those actually observed, nor even by those observed at some time, whether past, present, or future. The whole of the next chapter will be devoted to the explication of this term, which is intimately bound up with our conception of the structure of a scientific theory.

The distinction I have drawn between realism and anti-realism, in so far as it pertains to acceptance, concerns only how much belief is involved therein. Acceptance of theories (whether full, tentative, to a degree, etc.) is a phenomenon of scientific activity which clearly involves more than belief. One main reason for this is that we are never confronted with a complete theory. So if a scientist accepts a theory, he thereby involves himself in a certain sort of research programme. That programme could well be different from the one acceptance of another theory would have given him, even if those two (very incomplete) theories are equivalent to each other with respect to everything that is observable—in so far as they go.

Thus acceptance involves not only belief but a certain commitment. Even for those of us who are not working scientists, the acceptance involves a commitment to confront any future phenomena by means of the conceptual resources of this theory. It determines the terms in which we shall seek explanations. If the acceptance is at all strong, it is exhibited in the person's assumption of the role of explainer, in his willingness to answer questions *ex cathedra.* Even if you do not accept a theory, you can engage in discourse in a context in which language use is guided by that theory—but acceptance produces such contexts. There are similarities in all of this to ideological commitment. A commitment is of course not true or false: The confidence exhibited is that it will be *vindicated*.

This is a preliminary sketch of the *pragmatic* dimension of theory acceptance. Unlike the epistemic dimension, it does not figure overtly in the disagreement between realist and anti-realist. But because the amount of belief involved in acceptance is typically less according to anti-realists, they will tend to make more of the pragmatic aspects. It is as well to note here the important difference. Belief that a theory is true, or that it is empirically adequate, does not imply, and is not implied by, belief that full acceptance of the theory will be vindicated. To see this, you need only consider here a person who has quite definite beliefs about the future of the human race, or about the scientific community and the influences thereon and practical limitations we have. It might well be, for instance, that a theory which is empirically adequate will not combine easily with some other theories which we have accepted in fact, or that Armageddon will occur before we succeed. Whether belief that a theory is true, or that it is empirically adequate, can be equated with belief that acceptance of it would, under ideal research conditions, be vindicated in the long run, is another question. It seems to me an irrelevant question within philosophy of science, because an affirmative answer would not obliterate the distinction we have already established by the preceding remarks. (The question may also assume that counterfactual statements are objectively true or false, which I would deny.)

Although it seems to me that realists and anti-realists need not disagree about the pragmatic aspects of theory acceptance, I have mentioned it here because I think that typically they do. We shall find ourselves returning time and again, for example, to requests for explanation to which realists typically attach an objective validity which anti-realists cannot grant.

§2. *The Theory/Observation 'Dichotomy'*

For good reasons, logical positivism dominated the philosophy of science for thirty years. In 1960, the first volume of *Minnesota Studies in the Philosophy of Science* published Rudolf Carnap's 'The Methodological Status of Theoretical Concepts', which is, in many ways, the culmination of the positivist programme. It interprets science by relating it to an observation language (a postulated part of natural language which is devoid of theoretical terms). Two years later this article was followed in the same series by Grover Maxwell's 'The Ontological Status of Theoretical Entities', in title and theme a direct counter to Carnap's. This is the *locus classicus* for the new realists' contention that the theory/observation distinction cannot be drawn.

I shall examine some of Maxwell's points directly, but first a general remark about the issue. Such expressions as 'theoretical entity' and 'observable–theoretical dichotomy' are, on the face of it, examples of category mistakes. Terms or concepts are theoretical (introduced or adapted for the purposes of theory construction); entities are observable or unobservable. This may seem a little point, but it separates the discussion into two issues. Can we divide our language into a theoretical and non-theoretical part? On the other hand, can we classify objects and events into observable and unobservable ones?

Maxwell answers both questions in the negative, while not distinguishing them too carefully. On the first, where he can draw on well-known supportive essays by Wilfrid Sellars and Paul Feyerabend, I am in total agreement. All our language is thoroughly theory-infected. If we could cleanse our language of theory-laden terms, beginning with the recently introduced ones like 'VHF receiver', continuing through 'mass' and 'impulse' to 'element' and so on into the prehistory of language formation, we would end up with nothing useful. The way we talk, and scientists talk, is guided by the pictures provided by previously accepted theories. This is true also, as Duhem already emphasized, of experimental reports. Hygienic reconstructions of language such as the positivists envisaged are simply not on. I shall return to this criticism of positivism in the next chapter.

But does this mean that we must be scientific realists? We surely have more tolerance of ambiguity than that. The fact that we let our language be guided by a given picture, at some point, does not show how much we believe about that picture. When we speak of the sun coming up in the morning and setting at night, we are guided by a picture now explicitly disavowed. When Milton wrote *Paradise Lost* he deliberately let the old geocentric astronomy guide his poem, although various remarks in passing clearly reveal his interest in the new astronomical discoveries and speculations of his time. These are extreme examples, but show that no immediate conclusions can be drawn from the theory-ladenness of our language.

However, Maxwell's main arguments are directed against the observable–unobservable distinction. Let us first be clear on what this distinction was supposed to be. The term 'observable' classifies putative entities (entities which may or may not exist). A flying horse is observable—that is why we are so sure that there aren't any—and the number seventeen is not. There is supposed to be a correlate classification of human acts: an unaided act of perception, for instance, is an observation. A calculation of the mass of a particle from the deflection of its trajectory in a known force field, is not an observation of that mass.

It is also important here not to confuse *observing* (an entity, such as a thing, event, or process) and *observing that* (something or other is the case). Suppose one of the Stone Age people recently found in the Philippines is shown a tennis ball or a car crash. From his behaviour, we see that he has noticed them; for example, he picks up the ball and throws it. But he has not seen *that* it is a tennis ball, or *that* some event is a car crash, for he does not even have those concepts. He cannot get that information through perception; he would first have to learn a great deal. To say that he does not see the same things and events as we do, however, is just silly; it is a pun which trades on the ambiguity between seeing and seeing that. (The truth-conditions for our statement '*x* observes *that A*' must be such that what concepts *x* has, presumably related to the language *x* speaks if he is human, enter as a variable into the correct truth definition, in some way. To say that *x* observed the tennis ball, therefore, does not imply at all that *x* observed that it was a tennis ball; that would require some conceptual awareness of the game of tennis.)

The arguments Maxwell gives about observability are of two sorts: one directed against the possibility of drawing such distinctions, the other against the importance that could attach to distinctions that can be drawn.

The first argument is from the continuum of cases that lie between direct observation and inference:

there is, in principle, a continuous series beginning with looking through a vacuum and containing these as members: looking through a windowpane, looking through glasses, looking through binoculars, looking through a low-power microscope, looking through a high-power microscope, etc., in the order given. The important consequence is that, so far, we are left without criteria which would enable us to draw a non-arbitrary line between 'observation' and 'theory'.[4]

This continuous series of supposed acts of observation does not correspond directly to a continuum in what is supposed observable. For if something can be seen through a window, it can also be seen with the window raised. Similarly, the moons of Jupiter can be seen through a telescope; but they can also be seen without a telescope if you are close enough. That something is observable does not automatically imply that the conditions are right for observing it now. The principle is:

X is observable if there are circumstances which are such that, if X is present to us under those circumstances, then we observe it.

This is not meant as a definition, but only as a rough guide to the avoidance of fallacies.

We may still be able to find a continuum in what is supposed detectable: perhaps some things can only be detected with the aid of an optical microscope, at least; perhaps some require an electron microscope, and so on. Maxwell's problem is: where shall we draw the line between what is observable and what is only detectable in some more roundabout way?

Granted that we cannot answer this question without arbitrariness, what follows? That 'observable' is a *vague predicate*. There are many puzzles about vague predicates, and many sophisms designed to show that, in the presence of vagueness, no distinction can be drawn at all. In Sextus Empiricus, we find the argument that incest is not immoral, for touching your mother's big toe with your little finger is not immoral, and all the rest differs only by degree. But predicates in natural language are almost all vague, and there is no problem in their use; only in formulating the logic that governs them.[5] A vague predicate is usable provided it has clear cases and clear counter-cases. Seeing with the unaided eye is a clear case of observation. Is Maxwell then perhaps challenging us to present a clear counter-case? Perhaps so, for he says 'I have been trying to support the thesis that any (non-logical) term is a *possible* candidate for an observation term.'

A look through a telescope at the moons of Jupiter seems to me a clear case of observation, since astronauts will no doubt be able to see them as well from close up. But the purported observation of micro-particles in a cloud chamber seems to me a clearly different case—if our theory about what happens there is right. The theory says that if a charged particle traverses a chamber filled with saturated vapour, some atoms in the neighbourhood of its path are ionized. If this vapour is decompressed, and hence becomes supersaturated, it condenses in droplets on the ions, thus marking the path of the particle. The resulting silver-grey line is similar (physically as well as in appearance) to the vapour trail left in the sky when a jet passes. Suppose I point to such a trail and say: 'Look, there is a jet!'; might you not say: 'I see the vapour trail, but where is the jet?' Then I would answer: 'Look just a bit ahead of the trail ... there! Do you see it?' Now, in the case of the cloud chamber this response is not possible. So while the particle is detected by means of the cloud chamber, and the detection is based on observation, it is clearly not a case of the particle's being observed.

As a second argument, Maxwell directs our attention to the 'can' in 'what is observable is what can be observed.' An object might of course be temporarily unobservable—in a rather different sense: it cannot be observed in the circumstances in which it actually is at the moment, but could be observed if the circumstances were more favourable. In just the same way, I might be temporarily invulnerable or invisible. So we should concentrate on 'observable' *tout court*, or on (as he prefers to say) 'unobservable in principle'. This Maxwell explains as meaning that the relevant scientific theory *entails* that the entities cannot be observed in any circumstances. But this never happens, he says, because the different circumstances could be ones in which we have different sense organs—electron–microscope eyes, for instance.

This strikes me as a trick, a change in the subject of discussion. I have a mortar and pestle made of copper and weighing about a kilo. Should I call it breakable because a giant could break it? Should I call the Empire State Building portable? Is there no distinction between a portable and a console record player? The human organism is, from the point of view of physics, a certain kind of measuring apparatus. As such it has certain inherent limitations— which will be described in detail in the final physics and biology. It is these limitations to which the 'able' in 'observable' refers—our limitations, *qua* human beings.

As I mentioned, however, Maxwell's article also contains a different sort of argument: even if there is a feasible observable/

unobservable distinction, this distinction has no importance. The point at issue for the realist is, after all, the reality of the entities postulated in science. Suppose that these entities could be classified into observables and others; what relevance should that have to the question of their existence?

Logically, none. For the term 'observable' classifies putative entities, and has logically nothing to do with existence. But Maxwell must have more in mind when he says: 'I conclude that the drawing of the observational–theoretical line at any given point is an accident and a function of our physiological make-up, . . . and, there-fore, that it has no ontological significance whatever.'[6] No onto-logical significance if the question is only whether 'observable' and 'exists' imply each other—for they do not; but significance for the question of scientific realism?

Recall that I defined scientific realism in terms of the aim of science, and epistemic attitudes. The question is what aim scientific activity has, and how much we shall believe when we accept a scientific theory. What is the proper form of acceptance: belief that the theory, as a whole, is true; or something else? To this question, what is observable by us seems eminently relevant. Indeed, we may attempt an answer at this point: to accept a theory is (for us) to believe that it is empirically adequate—that what the theory says *about what is observable* (by us) is true.

It will be objected at once that, on this proposal, what the anti-realist decides to believe about the world will depend in part on what he believes to be his, or rather the epistemic community's, acces-sible range of evidence. At present, we count the human race as the epistemic community to which we belong; but this race may mutate, or that community may be increased by adding other animals (terrestrial or extra-terrestrial) through relevant ideological or moral decisions ('to count them as persons'). Hence the anti-realist would, on my proposal, have to accept conditions of the form

If the epistemic community changes in fashion Y, then my beliefs about the world will change in manner Z.

To see this as an objection to anti-realism is to voice the require-ment that our epistemic policies should give the same results independent of our beliefs about the range of evidence accessible to us. That requirement seems to me in no way rationally compelling; it could be honoured, I should think, only through a thorough-going scepticism or through a commitment to wholesale leaps of faith. But we cannot settle the major questions of epistemology *en passant* in philosophy of science; so I shall just conclude that it is, on the face of it, not irrational to commit oneself only to a search for theories that are empirically adequate, ones whose models fit the observable phenomena, while recognizing that what counts as an observable phenomenon is a function of what the epistemic community is (that *observable* is *observable-to-us*).

The notion of empirical adequacy in this answer will have to be spelt out very carefully if it is not to bite the dust among hackneyed objections. I shall try to do so in the next chapter. But the point stands: even if observability has nothing to do with existence (is, indeed, too anthropocentric for that), it may still have much to do with the proper epistemic attitude to science.

⚹ §3. *Inference to the Best Explanation* ⚹

A view advanced in different ways by Wilfrid Sellars, J. J. C. Smart, and Gilbert Harman is that the canons of rational inference require scientific realism. If we are to follow the same patterns of inference with respect to this issue as we do in science itself, we shall find ourselves irrational unless we assert the truth of the scientific theories we accept. Thus Sellars says: 'As I see it, to have good reason for holding a theory is *ipso facto* to have good reason for holding that the entities postulated by the theory exist.'[7]

The main rule of inference invoked in arguments of this sort is the rule of *inference to the best explanation*. The idea is perhaps to be credited to C. S. Peirce,[8] but the main recent attempts to explain this rule and its uses have been made by Gilbert Harman.[9] I shall only present a simplified version. Let us suppose that we have evidence *E*, and are considering several hypotheses, say *H* and *H'*. The rule then says that we should infer *H* rather than *H'* exactly if *H* is a better explanation of *E* than *H'* is. (Various qualifications are necessary to avoid inconsistency: we should always try to move to the best over-all explanation of all available evidence.)

It is argued that we follow this rule in all 'ordinary' cases; and that if we follow it consistently everywhere, we shall be led to scientific realism, in the way Sellars's dictum suggests. And surely there are many telling 'ordinary' cases: I hear scratching in the wall, the patter of little feet at midnight, my cheese disappears—and I infer that a mouse has come to live with me. Not merely that these apparent signs of mousely presence will continue, not merely that all the observable phenomena will be as if there is a mouse; but that there really is a mouse.

Will this pattern of inference also lead us to belief in unobservable entities? Is the scientific realist simply someone who consistently follows the rules of inference that we all follow in more mundane contexts? I have two objections to the idea that this is so.

First of all, what is meant by saying that we all *follow* a certain rule of inference? One meaning might be that we deliberately and consciously 'apply' the rule, like a student doing a logic exercise. That meaning is much too literalistic and restrictive; surely all of mankind follows the rules of logic much of the time, while only a fraction can even formulate them. A second meaning is that we act in accordance with the rules in a sense that does not require conscious deliberation. That is not so easy to make precise, since each logical rule is a rule of permission (*modus ponens* allows you to infer *B* from *A* and (if *A* then *B*), but does not forbid you to infer (*B or A*) instead). However, we might say that a person behaved in accordance with a set of rules in that sense if every conclusion he drew could be reached from his premisses via those rules. But this meaning is much too loose; in this sense we always behave in accordance with the rule that any conclusion may be inferred from any premiss. So it seems that to be following a rule, I must be willing to believe all conclusions it allows, while definitely unwilling to believe conclusions at variance with the ones it allows—or else, change my willingness to believe the premisses in question.

Therefore the statement that we all follow a certain rule in certain cases, is a *psychological hypothesis* about what we are willing and unwilling to do. It is an empirical hypothesis, to be confronted with data, and with rival hypotheses. Here is a rival hypothesis: we are always willing to believe that the theory which best explains the evidence, is empirically adequate (that all the observable phenomena are as the theory says they are).

In this way I can certainly account for the many instances in which a scientist appears to argue for the acceptance of a theory or hypothesis, on the basis of its explanatory success. (A number of such instances are related by Thagard.[8]) For, remember: I equate the acceptance of a scientific theory with the belief that it is empirically adequate. We have therefore two rival hypotheses concerning these instances of scientific inference, and the one is apt in a realist account, the other in an anti-realist account.

Cases like the mouse in the wainscoting cannot provide telling evidence between those rival hypotheses. For the mouse *is* an observable thing; therefore 'there is a mouse in the wainscoting' and 'All observable phenomena are as if there is a mouse in the wainscoting' are totally equivalent; each implies the other (given what we know about mice).

It will be countered that it is less interesting to know whether people do follow a rule of inference than whether they ought to follow it. Granted; but the premiss that we all follow the rule of inference to the best explanation when it comes to mice and other mundane matters—that premiss is shown wanting. It is not warranted by the evidence, because that evidence is not telling *for* the premiss *as against* the alternative hypothesis I proposed, which is a relevant one in this context.

My second objection is that even if we were to grant the correctness (or worthiness) of the rule of inference to the best explanation, the realist needs some further premiss for his argument. For this rule is only one that dictates a choice when given a set of rival hypotheses. In other words, we need to be committed to belief in one of a range of hypotheses before the rule can be applied. Then, under favourable circumstances, it will tell us which of the hypotheses in that range to choose. The realist asks us to choose between different hypotheses that explain the regularities in certain ways; but his opponent always wishes to choose among hypotheses of the form 'theory T_i is empirically adequate'. So the realist will need his special extra premiss that every universal regularity in nature needs an explanation, before the rule will make realists of us all. And that is just the premiss that distinguishes the realist from his opponents (and which I shall examine in more detail in Sections 4 and 5 below).

The logically minded may think that the extra premiss can be bypassed by logical *léger-de-main*. For suppose the data are that all facts observed so far accord with theory T; then T is one possible explanation of those data. A rival is *not-T* (that T is false). This rival is a very poor explanation of the data. So we *always* have a set of rival hypotheses, and the rule of inference to the best explanation leads us unerringly to the conclusion that T is true. Surely I am committed to the view that T is true or T is false?

This sort of epistemological rope-trick does not work of course. To begin, I am committed to the view that T is true or T is false, but not thereby committed to an inferential move to one of the two! The rule operates only if I have decided not to remain neutral between these two possibilities.

Secondly, it is not at all likely that the rule will be applicable to such logically concocted rivals. Harman lists various criteria to apply to the evaluation of hypotheses *qua* explanations.[10] Some are rather vague, like simplicity (but is simplicity not a reason to use a theory whether you believe it or not?). The precise ones come from statistical theory which has lately proved of wonderful use to epistemology:

H is a better explanation than H' (*ceteris paribus*) of E, provided:
  (a) $P(H) > P(H')$—H has higher probability than H'
  (b) $P(E/H) > P(E/H')$—H bestows higher probability on E than H' does.

The use of 'initial' or *a priori* probabilities in (a)—the initial plausibility of the hypotheses themselves—is typical of the so-called *Bayesians*. More traditional statistical practice suggests only the use of (b). But even that supposes that H and H' bestow definite probabilities on E. If H' is simply the denial of H, that is not generally the case. (Imagine that H says that the probability of E equals $\frac{3}{4}$. The very most that *not-H* will entail is that the probability of E is some number other than $\frac{3}{4}$; and usually it will not even entail that much, since H will have other implications as well.)

Bayesians tend to cut through this 'unavailability of probabilities' problem by hypothesizing that everyone has a specific subjective probability (degree of belief) for every proposition he can formulate. In that case, no matter what E, H, H' are, all these probabilities really are (in principle) available. But they obtain this availability by making the probabilities thoroughly subjective. I do not think that scientific realists wish their conclusions to hinge on the subjectively established initial plausibility of there being unobservable entities, so I doubt that this sort of Bayesian move would help here. (This point will come up again in a more concrete form in connection with an argument by Hilary Putnam.)

I have kept this discussion quite abstract; but more concrete arguments by Sellars, Smart, and Putnam will be examined below. It should at least be clear that there is no open-and-shut argument from common sense to the unobservable. Merely following the ordinary patterns of inference in science does not obviously and automatically make realists of us all.

§4. *Limits of the Demand for Explanation*

In this section and the next two, I shall examine arguments for realism that point to explanatory power as a criterion for theory choice. That this is indeed a criterion I do not deny. But these arguments for realism succeed only if the demand for explanation is supreme—if the task of science is unfinished, *ipso facto*, as long as any pervasive regularity is left unexplained. I shall object to this line of argument, as found in the writings of Smart, Reichenbach, Salmon, and Sellars, by arguing that such an unlimited demand for explanation leads to a demand for hidden variables, which runs contrary to at least one major school of thought in twentieth-century physics. I do not think that even these philosophers themselves wish to saddle realism with logical links to such consequences: but realist yearnings were born among the mistaken ideals of traditional metaphysics.

In his book *Between Science and Philosophy*, Smart gives two main arguments for realism. One is that only realism can respect the important distinction between *correct* and *merely useful* theories. He calls 'instrumentalist' any view that locates the importance of theories in their use, which requires only empirical adequacy, and not truth. But how can the instrumentalist explain the usefulness of his theories?

Consider a man (in the sixteenth century) who is a realist about the Copernican hypothesis but instrumentalist about the Ptolemaic one. He can explain the instrumental usefulness of the Ptolemaic system of epicycles because he can prove that the Ptolemaic system can produce almost the same predictions about the apparent motions of the planets as does the Copernican hypothesis. Hence the assumption of the realist truth of the Copernican hypothesis explains the instrumental usefulness of the Ptolemaic one. Such an explanation of the instrumental usefulness of certain theories would not be possible if *all* theories were regarded as merely instrumental.[11]

What exactly is meant by 'such an explanation' in the last sentence? If no theory is assumed to be true, then no theory has its usefulness explained as following from the truth of another one—granted. But would we have less of an explanation of the usefulness of the

Ptolemaic hypothesis if we began instead with the premiss that the Copernican gives implicitly a very accurate description of the motions of the planets as observed from earth? This would not assume the truth of Copernicus's heliocentric hypothesis, but would still entail that Ptolemy's simpler description was also a close approximation of those motions.

However, Smart would no doubt retort that such a response pushes the question only one step back: what explains the accuracy of predictions based on Copernicus's theory? If I say, the empirical adequacy of that theory, I have merely given a verbal explanation. For of course Smart does not mean to limit his question to actual predictions—it really concerns all actual and possible predictions and retrodictions. To put it quite concretely: what explains the fact that all observable planetary phenomena fit Copernicus's theory (if they do)? From the medieval debates, we recall the nominalist response that the basic regularities are merely brute regularities, and have no explanation. So here the anti-realist must similarly say: that the observable phenomena exhibit these regularities, because of which they fit the theory, is merely a brute fact, and may or may not have an explanation in terms of unobservable facts 'behind the phenomena'—it really does not matter to the goodness of the theory, nor to our understanding of the world.

Smart's main line of argument is addressed to exactly this point. In the same chapter he argues as follows. Suppose that we have a theory T which postulates micro-structure directly, and macro-structure indirectly. The statistical and approximate laws about macroscopic phenomena are only partially spelt out perhaps, and in any case derive from the precise (deterministic or statistical) laws about the basic entities. We now consider theory T', which is part of T, and says only what T says about the macroscopic phenomena. (How T' should be characterized I shall leave open, for that does not affect the argument here.) Then he continues:

I would suggest that the realist could (say) ... that the success of T' is explained by the fact that the original theory T is true of the things that it is ostensibly about; in other words by the fact that there really are electrons or whatever is postulated by the theory T. If there were no such things, and if T were not true in a realist way, would not the success of T' be quite inexplicable? One would have to suppose that there were innumerable lucky accidents about the behaviour mentioned in the observational vocabulary, so that they behaved miraculously *as if* they were brought about by the non-existent things ostensibly talked about in the theoretical vocabulary.[12]

In other passages, Smart speaks similarly of 'cosmic coincidences'. The regularities in the observable phenomena must be explained in terms of deeper structure, for otherwise we are left with a belief in lucky accidents and coincidences on a cosmic scale.

I submit that if the demand for explanation implicit in these passages were precisely formulated, it would at once lead to absurdity. For if the mere fact of postulating regularities, without explanation, makes T' a poor theory, T will do no better. If, on the other hand, there is some precise limitation on what sorts of regularities can be postulated as basic, the context of the argument provides no reason to think that T' must automatically fare worse than T.

In any case, it seems to me that it is illegitimate to equate being a lucky accident, or a coincidence, with having no explanation. It was by coincidence that I met my friend in the market—but I can explain why I was there, and he can explain why he came, so together we can explain how this meeting happened. We call it a coincidence, not because the occurrence was inexplicable, but because we did not severally go to the market in order to meet.[13] There cannot be a requirement upon science to provide a theoretical elimination of coincidences, or accidental correlations in general, for that does not even make sense. There is nothing here to motivate the demand for explanation, only a restatement in persuasive terms.

§5. *The Principle of the Common Cause*

Arguing against Smart, I said that if the demand for explanation implicit in his arguments were precisely formulated, it would lead to absurdity. I shall now look at a precise formulation of the demand for explanation: Reichenbach's principle of the common cause. As Salmon has recently pointed out, if this principle is imposed as a demand on our account of what there is in the world, then we are led to postulate the existence of unobservable events and processes.[14]

I will first state the argument, and Reichenbach's principle, in a rough, intuitive form, and then look at its precise formulation. Suppose that two sorts of events are found to have a correlation. A simple example would be that one occurs whenever the other does; but the correlation may only be statistical. There is apparently a significant correlation between heavy cigarette-smoking and cancer, though merely a statistical one. Explaining such a correlation requires finding what Reichenbach called a *common cause*. But, the argument runs, there are often among observable events no common causes of given observable correlations. Therefore, scientific explanation often requires that there be certain unobservable events.

Reichenbach held it to be a principle of scientific methodology that every statistical correlation (at least, every positive dependence) must be explained through common causes. This means then that the very project of science will necessarily lead to the introduction of unobservable structure behind the phenomena. Scientific explanation will be impossible unless there are unobservable entities; but the aim of science is to provide scientific explanation; therefore, the aim of science can only be served if it is true that there are unobservable entities.

To examine this argument, we must first see how Reichenbach arrived at his notion of common causes and how he made it precise. I will then argue that his principle cannot be a general principle of science at all, and secondly, that the postulation of common causes (when it does occur) is also quite intelligible without scientific realism.

Reichenbach was one of the first philosophers to recognize the radical 'probabilistic turn' of modern physics. The classical ideal of science had been to find a method of description of the world so fine that it could yield deterministic laws for all processes. This means that, if such a description be given of the state of the world (or, more concretely, of a single isolated system) at time t, then its state at later time $t + d$ is uniquely determined. What Reichenbach argued very early on is that this ideal has a factual presupposition: it is not logically necessary that such a fine method of description exists, even in principle.[15] This view became generally accepted with the development of quantum mechanics.

So Reichenbach urged philosophers to abandon that classical ideal as the standard of completeness for a scientific theory. Yet it is clear that, if science does not seek for deterministic laws relating events to what happened before them, it does seek for *some* laws. And so Reichenbach proposed that the correct way to view science is as seeking for 'common causes' of a probabilistic or statistical sort.

We can make this precise using the language of probability theory. Let A and B be two events; we use P to designate their probability of occurrence. Thus $P(A)$ is the probability that A occurs and $P(A\&B)$ the probability that both A and B occur. In addition, we must consider the probability that *A* occurs *given that B* occurs. Clearly the probability of rain *given that* the sky is overcast, is higher than the probability of rain in general. We say that *B* is statistically relevant to *A* if the probability of *A given B*—written $P(A/B)$—is different from $P(A)$. If $P(A/B)$ is higher than $P(A)$, we say that there is a positive correlation. Provided *A* and *B* are events which have some positive likelihood of occurrence (i.e. $P(A)$, $P(B)$ are not zero), this is a symmetric relationship. The precise definitions are these:

(a) the probability of *A* given *B* is defined provided $P(B) \neq O$, and is

$$P(A/B) = \frac{P(A\&B)}{P(B)}$$

(b) *B* is statistically relevant to *A* exactly if $P(A/B) \neq P(A)$
(c) there is a positive correlation between *A* and *B* exactly if $P(A\&B) > P(A) \cdot P(B)$
(d) from (a) and (c) it follows that, if $P(A) \neq O$ and $P(B) \neq O$, then there is a positive correlation between *A* and *B* exactly if

$$P(A/B) > P(A),$$

and also if and only if

$$P(B/A) > P(B)$$

To say therefore that there is a positive correlation between cancer and heavy cigarette-smoking, is to say that the incidence of cancer among heavy cigarette-smokers is greater than it is in the general population. But because of the symmetry of *A* and *B* in (d), this statement by itself gives no reason to think that the smoking produces the cancer rather than the cancer producing the smoking, or both being produced by some other factor, or by several other factors, if any.

We are speaking here of facts relating to the same time. The cause we seek in the past: heavy smoking at one time is followed (with certain probabilities) by heavy smoking at a later time, and also by being cancerous at that later time. We have in this past event *C* really found the *common cause* of this present correlation if

$$P(A/B\&C) = P(A/C)$$

We may put this as follows: relative to the information that *C* has occurred, *A* and *B* are statistically independent. We can define the probability of an event X, whether by itself or conditional on another event Y, *relative to C* as follows:

(e) the probability relative to C is defined as   ·

$$P_c(X) = P(X/C)$$
$$P_c(X/Y) = P_c(X \& Y) \div P_c(Y)$$
$$= P(X/Y \& C)$$
provided $P_c(Y) \neq 0$, $P(C) \neq 0$

So to say that C is the common cause for the correlation between A and B is to say that, relative to C there is no such correlation. C explains the correlation, because we notice a correlation only as long as we do not take C into account.

Reichenbach's *Principle of the Common Cause* is that *every* relation of positive statistical relevance must be explained by statistical past common causes, in the above way.[16] To put it quite precisely and in Reichenbach's own terms:

If coincidences of two events A and B occur more frequently than would correspond to their independent occurrence, that is, if the events satisfy the relation

(1) $P(A \& B) > P(A) . P(B)$,

then there exists a common cause C for these events such that the fork ACB is *conjunctive*, that is, satisfies relations (2)–(5) below:

(2) $P(A \& B/C) = P(A/C) . P(B/C)$
(3) $P(A \& B/\overline{C}) = P(A/\overline{C}) . P(B/\overline{C})$
(4) $P(A/C) > P(A/\overline{C})$
(5) $P(B/C) > P(B/\overline{C})$

(1) follows logically from (2)–(5).

This principle of the common cause is at once precise and persuasive. It may be regarded as a formulation of the conviction that lies behind such arguments as that of Smart, requiring the elimination of 'cosmic coincidence' by science. But it is not a principle that guides twentieth-century science, because it is too close to the demand for deterministic theories of the world that Reichenbach wanted to reject. I shall show this by means of a schematic example; but this example will incorporate the sorts of non-classical correlations which distinguish quantum mechanics from classical physics. I refer here to the correlations exhibited by the thought experiment of Einstein, Podolski, and Rosen in their famous paper 'Can Quantum-Mechanical Description of Reality be Considered Complete?' These correlations are not merely theoretical: they are found in many actual experiments, such as Compton scattering and photon pair production. I maintain in addition that correlations sufficiently similar to refute the principle of the common cause must appear in almost any indeterministic theory of sufficient complexity.[17] Imagine that you have studied the behaviour of a system or object which, after being in state S, always goes into another state which may be characterized by various attributes $F_1, \ldots, F_n$ and $G_1, \ldots, G_n$. Suppose that you have come to the conclusion that this transition is genuinely indeterministic, but you can propose a theory about the transition probabilities:

(8) (a) $P(F_i/S) = 1/n$ (b) $P(G_i/S) = 1/n$
     (c) $P(F_i \equiv G_i/S) = 1$

where $\equiv$ means *if and only if* or *when and exactly when*. In other words, it is pure chance whether the state to which S transits is characterized by a given one of the F-attributes, and similarly for the G-attributes, but certain that it is characterized by F_1 if it is characterized by G_1, by F_2 if by G_2, and so on.

If we are convinced that this is an irreducible, indeterministic phenomenon, so that S is a complete description of the initial state, then we have a violation of the principle of the common cause. For from (8) we can deduce

(9) $P(F_i/S) . P(G_i/S) \quad = 1/n^2$
     $P(F_i \& G_i/S) = P(F_i/S) = 1/n$

which numbers are equal only if n is zero or one—the deterministic case. In all other cases, S does not qualify as the common cause of the new state's being F_i and G_i, and if S is complete, nothing else can qualify either.

The example I have given is schematic and simplified, and besides its indeterminism, it also exhibits a certain discontinuity, in that we discuss the transition of a system from one state S into a new state. In classical physics, if a physical quantity changed its value from i to j it would do so by taking on all the values between i and j in succession, that is, changing continuously. Would Reichenbach's principle be obeyed at least in some non-trivial, indeterministic theory in which all quantities have a continuous spectrum of values and all change is continuous? I think not, but I shall not argue this further. The question is really academic, for if the principle requires that, then it is also not acceptable to modern physical science.

Could one change a theory which violates Reichenbach's principle into one that obeys it, without upsetting its empirical adequacy? Possibly; one would have to deny that the attribution of state S gives complete information about the system at the time in question, and postulate *hidden parameters* that underlie these states. Attempts to do so for quantum mechanics are referred to as *hidden variable theories*, but it can be shown that if such a theory is empirically equivalent to orthodox quantum mechanics, then it still exhibits non-local correlations of a non-classical sort, which would still violate Reichenbach's principle. But again, the question is academic, since modern physics does not recognize the need for such hidden variables.

Could Reichenbach's principle be weakened so as to preserve its motivating spirit, while eliminating its present unacceptable consequences? As part of a larger theory of explanation (which I shall discuss later), Wesley Salmon has proposed to disjoin equation (2) above with

$$(2^*) \qquad P(A\&B/C) > P(A/C) \cdot P(B/C)$$

in which case C would still qualify as common cause. Note that in the schematic example I gave, S would then qualify as a common cause for the events F_i and G_j.

But so formulated, the principle yields a regress. For suppose (2^*) is true. Then we note a positive correlation *relative to C*:

$$P_c(A\&B) > P_c(A) \cdot P_c(B)$$

to which the principle applies and for which it demands a common cause C'. This regress stops only if, at some point, the exhibited common cause satisfies the original equation (2), which brings us back to our original situation; or if some other principle is used to curtail the demand for explanation.

In any case, weakening the principle in various ways (and certainly it will have to be weakened if it is going to be acceptable in any sense) will remove the force of the realist arguments. For any weakening is an agreement to leave some sorts of 'cosmic coincidence' unexplained. But that is to admit the tenability of the nominalist/empiricist point of view, for the demand for explanation ceases then to be a scientific 'categorical imperative'.

Nevertheless, there is a problem here that should be faced. Without a doubt, many scientific enterprises can be characterized as searches for common causes to explain correlations. What is the anti-realist to make of this? Are they not searches for explanatory realities behind the phenomena?

I think that there are two senses in which a principle of common causes is operative in the scientific enterprise, and both are perfectly intelligible without realism.

To the anti-realist, all scientific activity is ultimately aimed at greater knowledge of what is observable. So he can make sense of a search for common causes only if that search aids the acquisition of that sort of knowledge. But surely it does! When past heavy smoking is postulated as a causal factor for cancer, this suggests a further correlation between cancer and either irritation of the lungs, or the presence of such chemicals as nicotine in the bloodstream, or both. The postulate will be vindicated if such suggested further correlations are indeed found, and will, if so, have aided in the search for larger scale correlations among observable events.[18] This view reduces the Principle of Common Cause from a regulative principle for all scientific activity to one of its tactical maxims.

There is a second sense in which the principle of the common cause may be operative: as advice for the construction of theories and models. One way to construct a model for a set of observable correlations is to exhibit hidden variables with which the observed ones are individually correlated. This is a theoretical enterprise, requiring mathematical embedding or existence proofs. But if the resulting theory is then claimed to be empirically adequate, there is no claim that all aspects of the model correspond to 'elements of reality'. As a theoretical directive, or as a practical maxim, the principle of the common cause may well be operative in science—but not as a demand for explanation which would produce the metaphysical baggage of hidden parameters that carry no new empirical import.

§6. Limits to Explanation: a Thought Experiment

Wilfrid Sellars was one of the leaders of the return to realism in philosophy of science and has, in his writings of the past three decades, developed a systematic and coherent scientific realism. I have discussed a number of his views and arguments elsewhere; but will here concentrate on some aspects that are closely related to the arguments of Smart, Reichenbach, and Salmon just examined.[19] Let me begin by setting the stage in the way Sellars does.

There is a certain over-simplified picture of science, the 'levels picture', which pervades positivist writings and which Sellars successfully demolished.[20] In that picture, singular observable facts ('this crow is black') are scientifically explained by general observable regularities ('all crows are black') which in turn are explained by highly theoretical hypotheses not restricted in what they say to the observable. The three levels are commonly called those of *fact*, of *empirical law*, and of *theory*. But, as Sellars points out, theories do not explain, or even entail such empirical laws—they only show why observable things obey these so-called laws to the extent they do.[21] Indeed, perhaps we have no such empirical laws at all: all crows are black—except albinos; water boils at 100°C—provided atmospheric pressure is normal; a falling body accelerates—provided it is not intercepted, or attached to an aeroplane by a static line; and so forth. On the level of the observable we are liable to find only putative laws heavily subject to unwritten *ceteris paribus* qualifications.

This is, so far, only a methodological point. We do not really expect theories to 'save' our common everyday generalizations, for we ourselves have no confidence in their strict universality. But a theory which says that the micro-structure of things is subject to *some* exact, universal regularities, must imply the same for those things themselves. This, at least, is my reaction to the points so far. Sellars, however, sees an inherent inferiority in the description of the observable alone, an incompleteness which requires (*sub specie* the aims of science) an introduction of an unobservable reality behind the phenomena. This is brought out by an interesting 'thought-experiment'.

Imagine that at some early stage of chemistry it had been found that different samples of gold dissolve in *aqua regia* at different rates, although 'as far as can be observationally determined, the specimens and circumstances are identical'.[22] Imagine further that the response of chemistry to this problem was to postulate two distinct micro-

structures for the different samples of gold. Observationally un-
predictable variation in the rate of dissolution is explained by saying
that the samples are mixtures (not compounds) of these two
(observationally identical) substances, each of which has a fixed rate
of dissolution.

In this case we have explanation through laws which have no
observational counterparts that can play the same role. Indeed, no
explanation seems possible unless we agree to find our physical
variables outside the observable. But science aims to explain, must
try to explain, and so must require a belief in this unobservable
micro-structure. So Sellars contends.

There are at least three questions before us. Did this postulation
of micro-structure really have no new consequences for the observable phenomena? Is there really such a demand upon science that
it must explain—even if the means of explanation bring no gain in
empirical predictions? And thirdly, could a *different* rationale exist
for the use of a micro-structure picture in the development of a
scientific theory in a case like this?

First, it seems to me that these hypothetical chemists did postulate new observable regularities as well. Suppose the two substances
are A and B, with dissolving rates x and $x+y$ and that every gold
sample is a mixture of these substances. Then it follows that every
gold sample dissolves at a rate no lower than x and no higher than
$x+y$; *and* that between these two any value may be found—
to within the limits of accuracy of gold mixing. None of this is
implied by the data that different samples of gold have dissolved at
various rates between x and $x+y$. So Sellar's first contention is
false.

We may assume, for the sake of Sellars's example, that there is
still no way of predicting dissolving rates any further. Is there then
a categorical demand upon science to explain this variation which
does not depend on other observable factors? We have seen that a
precise version of such a demand (Reichenbach's principle of the
common cause) could result automatically in a demand for hidden
variables, providing a 'classical' underpinning for indeterministic
theories. Sellars recognized very well that a demand for hidden
variables would run counter to the main opinions current in
quantum physics. Accordingly he mentions '.... the familiar point
that the irreducibly and lawfully statistical ensembles of quantum-mechanical theory are mathematically inconsistent with the

assumption of hidden variables.'[23] Thus, he restricts the demand for explanation, in effect, to just those cases where it is *consistent* to add hidden variables to the theory. And consistency is surely a logical stopping-point.

This restriction unfortunately does not prevent the disaster. For while there are a number of proofs that hidden variables cannot be supplied so as to turn quantum mechanics into a classical sort of deterministic theory, those proofs are based on requirements much stronger than consistency. To give an example, one such assumption is that two distinct physical variables cannot have the same statistical distributions in measurement on all possible states.[24] Thus it is assumed that, if we cannot point to some possible difference in empirical predictions, then there is no real difference at all. If such requirements were lifted, and consistency alone were the criterion, hidden variables could indeed be introduced. I think we must conclude that science, in contrast to scientific realism, does not place an overriding value on explanation in the absence of any gain for empirical results.

Thirdly, then, let us consider how an anti-realist could make sense of those hypothetical chemists' procedure. After pointing to the new empirical implications which I mentioned three paragraphs ago, he would point to methodological reasons. By imagining a certain sort of micro-structure for gold and other metals, say, we might arrive at a theory governing many observationally disparate substances; and this might then have implications for new, wider empirical regularities when such substances interact. This would only be a hope, of course; no hypothesis is guaranteed to be fruitful—but the point is that the true demand on science is not for explanation *as such*, but for imaginative pictures which have a hope of suggesting new statements of observable regularities and of correcting old ones. This point is exactly the same as that for the principle of the common cause.

§7. *Demons and the Ultimate Argument*

Hilary Putnam, in the course of his discussions of realism in logic and mathematics, advanced several arguments for scientific realism as well. In *Philosophy of Logic* he concentrates largely on indispensability arguments—concepts of mathematical entities are indispensable to non-elementary mathematics, theoretical concepts are indispensable to physics.[25] Then he confronts the philosophical position of Fictionalism, which he gleans from the writings of Vaihinger and Duhem:

(T)he fictionalist says, in substance, 'Yes, certain concepts ... are indispensable, but no, that has no tendency to show that entities corresponding to those concepts actually exist. It only shows that those 'entities' are useful *fictions'*.[26]

Glossed in terms of theories: even if certain kinds of theories are indispensable for the advance of science, that does not show that those theories are true *in toto*, as well as empirically correct.

Putnam attacks this position in a roundabout way, first criticizing bad arguments against Fictionalism, and then garnering his reasons for rejecting Fictionalism from that discussion. The main bad reason he sees is that of Verificationism. The logical positivists adhered to the verificationist theory of meaning; which is roughly that the total cognitive content of an assertion, all that is meaningful in it, is a function of what empirical results would verify or refute it. Hence, they would say that there are no real differences between two hypotheses with the same empirical content. Consider two theories of what the world is like: Rutherford's atomic theory, and Vaihinger's hypothesis that, although perhaps there are no electrons and such, the observable world is nevertheless exactly as if Rutherford's theory were true. The Verificationist would say: these two theories, although Vaihinger's appears to be consistent with the denial of Rutherford's, amount to exactly the same thing.

Well, they don't, because the one says that there are electrons, and the other allows that there may not be. Even if the observable phenomena are as Rutherford says, the unobservable may be different. However, the positivists would say, if you argue that way, then you will automatically become a prey to scepticism. You will have to admit that there are possibilities you cannot prove or disprove by experiment, and so you will have to say that we just cannot know what the world is like. Worse; you will have no reason to reject any number of outlandish possibilities; demons, witchcraft, hidden powers collaborating to fantastic ends.

Putnam considers this argument for Verificationism to be mistaken, and his answer to it, strangely enough, will also yield an answer to the Fictionalism rejected by the verificationist. To dispel the bogey of scepticism, Putnam gives us a capsule introduction to contemporary (Bayesian) epistemology: Rationality requires that if

two hypotheses have all the same testable consequences (consequences for evidence that could be gathered), then we should not
accept the one which is *a priori the less plausible*. Where do we get
our *a priori* plausibility orderings? These we supply ourselves, either
individually or as communities: to accept a plausibility ordering is
neither

to make a judgment of empirical fact nor to state a theorem of deductive
logic; it is to take a methodological stand. One can only say whether the
demon hypothesis is 'crazy' or not if one has taken such a stand; I report
the stand I have taken (and, speaking as one who has taken this stand, I add:
and the stand all rational men take, implicitly or explicitly).[27]

On this view, the difference between Rutherford and Vaihinger, or
between Putnam and Duhem, is that (although they presumably
agree on the implausibility of demons) they disagree on the *a priori*
plausibility of electrons. Does each simply report the stand he has
taken, and add: this is, in my view, the stand of all rational men?
How disappointing.

Actually, it does not quite go that way. Putnam has skilfully
switched the discussion from electrons to demons, and asked us to
consider how we could rule out their existence. As presented,
however, Vaihinger's view differed from Rutherford's by being
logically weaker—it only withheld assent to an existence assertion.
It follows automatically that Vaihinger's view cannot be *a priori*
less plausible than Rutherford's. Putnam's ideological manœuvre
could at most be used to accuse an 'atheistic' anti-realist of irrationality (relative to Putnam's own stand, of course)—not one of the
agnostic variety.

Putnam concludes this line of reasoning by asking what more
could be wanted as evidence for the truth of a theory than what
the realist considers sufficient: 'But then . . . what *further* reasons
could one want before one regarded it as rational to *believe* a
theory?'[28] The answer is: *none*—at least if he equates reasons here
either with empirical evidence or with compelling arguments.
(Inclining reasons are perhaps another matter, especially because
Putnam uses the phrase 'rational to believe' rather than 'irrational
not to believe'.) Since Putnam has just done us the service of refuting
Verificationism, this answer 'none' cannot convict us of irrationality.
He has himself just argued forcefully that theories could agree in
empirical content and differ in truth-value. Hence, a realist will have

to make a leap of faith. The decision to leap is subject to rational scrutiny, but not *dictated* by reason and evidence.

In a further paper, 'What is Mathematical Truth', Putnam continues the discussion of scientific realism, and gives what I shall call the *Ultimate Argument*. He begins with a formulation of realism which he says he learned from Michael Dummett:

A realist (with respect to a given theory or discourse) holds that (1) the sentences of that theory are true or false; and (2) that what makes them true or false is something external—that is to say, it is not (in general) our sense data, actual or potential, or the structure of our minds, or our language, etc.[29]

This formulation is quite different from the one I have given even if we instantiate it to the case in which that theory or discourse is science or scientific discourse. Because the wide discussion of Dummett's views has given some currency to his usage of these terms, and because Putnam begins his discussion in this way, we need to look carefully at this formulation.

In my view, Dummett's usage is quite idiosyncratic. Putnam's statement, though very brief, is essentially accurate. In his 'Realism', Dummett begins by describing various sorts of realism in the traditional fashion, as disputes over whether there really exist entities of a particular type. But he says that in some cases he wishes to discuss, such as the reality of the past and intuitionism in mathematics, the central issues seem to him to be about other questions. For this reason he proposes a new usage: he will take such disputes

as relating, not to a class of entities or a class of terms, but to a class of *statements* ... Realism I characterize as the belief that statements of the disputed class possess an objective truth-value, independently of our means of knowing it: they are true or false in virtue of a reality existing independently of us. The anti-realist opposes to this the view that statements of the disputed class are to be understood only by reference to the sort of thing which we count as evidence for a statement of that class.[30]

Dummett himself notes at once that nominalists are realists in this sense.[31] If, for example, you say that abstract entities do not exist, and sets are abstract entities, hence sets do not exist, then you will certainly accord a truth-value to every statement of set theory. It might be objected that if you take this position then you have a decision procedure for determining the truth-values of these statements (*false* for existentially quantified ones, *true* for universal ones, apply truth tables for the rest). Does that not mean that, on your view, the truth-values are not independent of our knowledge? Not at all; for you clearly believe that if we had not existed, and *a fortiori* had had no knowledge, the state of affairs with respect to abstract entities would be the same.

Has Dummett perhaps only laid down a necessary condition for realism, in his definition, for the sake of generality? I do not think so. In discussions of quantum mechanics we come across the view that the particles of microphysics are real, and obey the principles of the theory, but at any time t when 'particle x has exact momentum p' is true then 'particle x has position q' is neither true nor false. In any traditional sense, this is a realist position with respect to quantum mechanics.

We note also that Dummett has, at least in this passage, taken no care to exclude non-literal construals of the theory, as long as they are truth–valued. The two are not the same; when Strawson construed 'The king of France in 1905 is bald' as neither true nor false, he was not giving a non-literal construal of our language. On the other hand, people tend to fall back on non-literal construals typically in order to be able to say, 'properly construed, the theory is true.'[32]

Perhaps Dummett is right in his assertion that what is really at stake, in realist disputes of various sorts, is questions about language—or, if not really at stake, at least the only serious philosophical problems in those neighbourhoods. Certainly the arguments in which he engages are profound, serious, and worthy of our attention. But it seems to me that his terminology ill accords with the traditional one. Certainly I wish to define scientific realism so that it need not imply that all statements in the theoretical language are true or false (only that they are all capable of being true or false, that is, there are conditions for each under which it has a truth-value); to imply nevertheless that the aim is that the theories should be true. And the contrary position of constructive empiricism is not anti-realist in Dummett's sense, since it also assumes scientific statements to have truth-conditions entirely independent of human activity or knowledge. But then, I do not conceive the dispute as being about language at all.

In any case Putnam himself does not stick with this weak formulation of Dummett's. A little later in the paper he directs himself to scientific realism *per se*, and formulates it in terms borrowed, he says, from Richard Boyd. The new formulation comes in the course of a new argument for scientific realism, which I shall call the Ultimate Argument:

> the positive argument for realism is that it is the only philosophy that doesn't make the success of science a miracle. That terms in mature scientific theories typically refer (this formulation is due to Richard Boyd), that the theories accepted in a mature science are typically approximately true, that the same term can refer to the same thing even when it occurs in different theories— these statements are viewed by the scientific realist not as necessary truths but as part of the only scientific explanation of the success of science, and hence as part of any adequate scientific description of science and its relations to its objects.[33]

Science, apparently, is required to explain its own success. There is this regularity in the world, that scientific predictions are regularly fulfilled; and this regularity, too, needs an explanation. Once *that* is supplied we may perhaps hope to have reached the *terminus de jure*?

The explanation provided is a very traditional one—*adequatio ad rem*, the 'adequacy' of the theory to its objects, a kind of mirroring of the structure of things by the structure of ideas—Aquinas would have felt quite at home with it.

Well, let us accept for now this demand for a scientific explanation of the success of science. Let us also resist construing it as merely a restatement of Smart's 'cosmic coincidence' argument, and view it instead as the question why we have successful scientific theories at all. Will this realist explanation with the Scholastic look be a scientifically acceptable answer? I would like to point out that science is a biological phenomenon, an activity by one kind of organism which facilitates its interaction with the environment. And this makes me think that a very different kind of scientific explanation is required.

I can best make the point by contrasting two accounts of the mouse who runs from its enemy, the cat. St. Augustine already remarked on this phenomenon, and provided an intentional explanation: the mouse *perceives that* the cat is its enemy, hence the mouse runs. What is postulated here is the 'adequacy' of the mouse's thought to the order of nature: the relation of enmity is correctly reflected in his mind. But the Darwinist says: Do not ask why the *mouse* runs from its enemy. Species which did not cope with their natural enemies no longer exist. That is why there are only ones who do.

In just the same way, I claim that the success of current scientific theories is no miracle. It is not even surprising to the scientific (Darwinist) mind. For any scientific theory is born into a life of fierce competition, a jungle red in tooth and claw. Only the successful theories survive—the ones which *in fact* latched on to actual regularities in nature.[34]

3

To Save The Phenomena[1]

> Physicists call a theory satisfactory if (1) it agrees with the experimental facts, (2) it is logically consistent, and (3) it is simple as compared to other explanations ... In fact, the author's interest in hidden-variable theories was kindled only when recently he became aware of the possibility of such experimental tests.
> On the other hand, we do not want to ignore the metaphysical implications of the theory.
>
> F. J. Belinfante, Foreword, *A Survey of Hidden-Variable Theories* (1973)

THE realist arguments discussed so far were developed mainly in a critique of logical positivism. Much of that critique was correct and successful: the positivist picture of science no longer seems tenable. Since that was essentially the only picture of science within philosophical ken, it is imperative to develop a new account of the structure of science. This account should especially provide a new answer to the question: what is the *empirical content* of a scientific theory?

§1. *Models*

Before turning to examples, let us distinguish the syntactic approach to theories from the semantic one which I favour. Modern axiomatics stems from the discussion of alternative geometric theories, which followed the development of non-Euclidean geometry in the nineteenth century. The first meta-mathematics was meta-geometry (a term already used in Bertrand Russell's *Essays on the Foundations of Geometry* in 1897). It will be easiest perhaps to introduce the relevant axiomatic concepts by way of some simple geometric theories. Consider the axioms:[2]

*A*0 There is at least one line.

*A*1 For any two lines, there is at most one point that lies on both.
*A*2 For any two points, there is exactly one line that lies on both.
*A*3 On every line there lie at least two points.
*A*4 There are only finitely many points.
*A*5 On any line there lie infinitely many points.

We have here the makings of three theories: T_0 has axioms *A*1–*A*3; T_1 is T_0 plus *A*4; and T_2 is T_0 plus *A*5.

There are some simple logical properties and relations easily observed here. Each of the three theories is *consistent*: no contradictions can be deduced. Secondly, T_1 and T_2 are *inconsistent with each other*: a contradiction can be deduced if we add *A*5 to T_1. Thirdly, T_1 and T_2 each *imply* T_0: all theorems of T_0 are clearly also theorems of the other two. The first achievement of modern symbolic logic was to give these logical properties and relations precise syntactic definitions, solely in terms of rules for manipulating symbols.

Yet, it will also be noticed that these logical notions have counterparts in relations expressible in terms of what the theory says, what it is about, and what it could be interpreted as being about. For instance, the consistency of theory T_1 is easiest to show by exhibiting a simple finite geometric structure of which axioms *A*1–*A*4 are true. This is the so-called Seven Point Geometry:

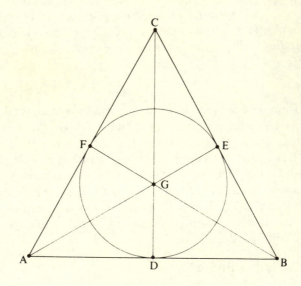

In this structure, only seven things are called 'points', namely, A, B, C, D, E, F, G. And equally, there are only seven 'lines', namely the three sides of the triangle, the three perpendiculars, and the inscribed circle. The first four axioms are easily seen to be true of this structure: the line DEF (i.e. the inscribed circle) has exactly three points on it, namely, D, E, and F; the points F and E have exactly one line lying on both, namely DEF; lines DEF and BEC have exactly one point in common, namely E; and so forth.

Any structure which satisfies the axioms of a theory in this way is called a *model* of that theory. (At the end of this section I shall relate this to other uses of the word 'model'.) Hence, the structure just exhibited is a model of T_1, and also of T_0, but not of T_2. The existence of a model establishes consistency by a very simple straight-forward argument:

> all the axioms of the theory (suitably interpreted) are true of the model; hence all the theorems are similarly true of it; but no contradiction can be true of anything; therefore, no theorem is a contradiction.

Thus logical claims, formulated in purely syntactic terms, can nevertheless often be demonstrated more simply by a detour via a look at models—but the notions of *truth* and *model* belong to semantics.

Nor is semantics merely the handmaiden of logic. For look at the theories T_1 and T_2; logic tells us that these are inconsistent with each other, and there is an end to it. The axioms of T_1 can only be satisfied by finite structures; the axioms of T_2, however, are satisfied only by infinite ones such as the Euclidean plane.

Yet you will have noticed that I drew a Euclidean triangle to convey what the Seven Point Geometry looks like. For that seven-point structure can be *embedded* in a Euclidean structure. We say that one structure can be embedded in another, if the first is iso-morphic to a part (substructure) of the second. Isomorphism is of course total identity of structure and is a limiting case of embedda-bility: if two structures are isomorphic then each can be embedded in the other. The seven-point geometry is isomorphic to a certain Euclidean plane figure, or in other words, it can be embedded in the Euclidean plane. This points to a much more interesting relationship between the theories T_1 and T_2 than inconsistency:

> every model of T_1 can be embedded in (identified with a sub-structure of) a model of T_2.

This sort of relationship, which is peculiarly semantic, is clearly very important for the comparison and evaluation of theories, and is not accessible to the syntactic approach.

The syntactic picture of a theory identifies it with a body of theorems, stated in one particular language chosen for the expression of that theory. This should be contrasted with the alternative of presenting a theory in the first instance by identifying a class of structures as its models. In this second, semantic, approach the language used to express the theory is neither basic nor unique; the same class of structures could well be described in radically different ways, each with its own limitations. The models occupy centre stage.

The use of the word 'model' in this discussion derives from logic and meta-mathematics. Scientists too speak of models, and even of models of a theory, and their usage is somewhat different. 'The Bohr model of the atom', for example, does not refer to a single structure. It refers rather to a type of structure, or class of structures, all sharing certain general characteristics. For in that usage, the Bohr model was intended to fit hydrogen atoms, helium atoms, and so forth. Thus in the scientists' use, 'model' denotes what I would call a model-type. Whenever certain parameters are left unspecified in the description of a structure, it would be more accurate to say (contrary of course to common usage and convenience) that we described a structure-type. Nevertheless, the usages of 'model' in meta-mathematics and in the sciences are not as far apart as has sometimes been said. I will continue to use the word 'model' to refer to specific structures, in which all relevant parameters have specific values.

Rather than pursue this general discussion I turn now to a concrete example of a physical theory, in order to introduce the crucially relevant notions by illustration.

§2. *Apparent Motion and Absolute Space*

When Newton wrote his *Mathematical Principles of Natural Philosophy* and *System of the World*, he carefully distinguished the phenomena to be saved from the reality to be postulated. He distinguished the 'absolute magnitudes' which appear in his axioms from their 'sensible measures' which are determined experimentally. He discussed carefully the ways in which, and extent to which, 'the true motions of particular bodies may be determined from the apparent', via the assertion that 'the apparent motions ... are the differences of true motions'.[3]

We can illustrate these distinctions through the discussion of planetary motion preceding Newton. Ptolemy described these motions on the assumption that the earth was stationary. For him, there was no distinction between true and apparent motion: the true motion is exactly what is seen in the heavens. (What that motion is, may of course not be evident at once: it takes thought to realize that a planet's motion really does look like a circular motion around a moving centre.) In Copernicus's theory, the sun is stationary. Hence, what we see is only the planets' motion relative to the earth, which is not itself stationary. The apparent motion of the planets is identified as the difference between the earth's true motion and the planets' true motion—true motion being, in this case, motion relative to the sun. Finally, Newton, in his general mechanics, did not assume that either the earth or the sun is stationary. And he generalized the idea of apparent motion—which is motion relative to the earth—to that of motion of one body relative to another. We can speak of the planets' motion relative to the sun, or to the earth, or to the moon, or what have you. What is observed is always some relative motion: an apparent motion is a motion relative to the observer. And Newton held that relative motions are always identifiable as a difference between true motions, whatever those may be (an assertion which can be given precise content using vector representation of motion).

The 'apparent motions' form relational structures defined by measuring relative distances, time intervals, and angles of separation. For brevity, let us call these relational structures *appearances*. In the mathematical model provided by Newton's theory, bodies are located in Absolute Space, in which they have real or absolute motions. But within these models we can define structures that are meant to be exact reflections of those appearances, and are, as Newton says, identifiable as differences between true motions. These structures, defined in terms of the relevant relations between absolute locations and absolute times, which are the appropriate parts of Newton's models, I shall call *motions*, borrowing Simon's term.[4] (Later I shall use the more general term *empirical substructures*.)

When Newton claims empirical adequacy for his theory, he is claiming that his theory has some model such that *all actual appearances are identifiable with (isomorphic to) motions* in that model. (This refers of course to all actual appearances throughout the history of the universe, and whether in fact observed or not.)

Newton's theory does a great deal more than this. It is part of his theory that there is such a thing as Absolute Space, that absolute motion is motion relative to Absolute Space, that absolute acceleration causes certain stresses and strains and thereby deformations in the appearances, and so on. He offered in addition the *hypothesis* (his term) that the centre of gravity of the solar system is at rest in Absolute Space.[5] But as he himself noted, the appearances would be no different if that centre were in any other state of constant absolute motion. This is the case for two reasons: differences between true motions are not changed if we add a constant factor to all velocities; and force is related to changes in motion (accelerations) and not to motion directly.

Let us call Newton's theory (mechanics and gravitation) TN, and $TN(v)$ the theory TN plus the postulate that the centre of gravity of the solar system has constant absolute velocity v. By Newton's own account, he claims empirical adequacy for $TN(0)$; and also that, if $TN(0)$ is empirically adequate, then so are all the theories $TN(v)$.

Recalling what it was to claim empirical adequacy, we see that all the theories $TN(v)$ are empirically equivalent exactly *if all the motions in a model of $TN(v)$ are isomorphic to motions in a model of $TN(v+w)$*, for all constant velocities v and w. For now, let us agree that these theories are empirically equivalent, referring objections to a later section.

§3. *Empirical Content of Newton's Theory*

What exactly is the 'empirical import' of $TN(0)$? Let us focus on a fictitious and anachronistic philosopher, Leibniz*, whose only quarrel with Newton's theory is that he does not believe in the existence of Absolute Space. As a corollary, of course, he can attach no 'physical significance' to statements about absolute motion. Leibniz* believes, like Newton, that $TN(0)$ is empirically adequate; but not that it is true. For the sake of brevity, let us say that Leibniz* *accepts* the theory but that he does not *believe* it; when confusion threatens we may expand that idiom to say that he *accepts the theory as empirically adequate*, but does not *believe it to be true*. What does Leibniz* believe, then?

Leibniz* believes that $TN(0)$ is empirically adequate, and hence equivalently, that all the theories $TN(v)$ are empirically adequate. Yet we cannot identify the theory which Leibniz* holds about the

world—call it *TNE*—with the common part of all the theories
TN(v). For each of the theories *TN(v)* has such consequences as that
the earth has *some* absolute velocity, and that Absolute Space exists.
In each model of each theory *TN(v)* there is to be found something
other than motions, and there is the rub.

To believe a theory is to believe that one of its models correctly
represents the world. You can think of the models as representing
the possible worlds allowed by the theory; one of these possible
worlds is meant to be the real one. To believe the theory is to
believe that exactly one of its models correctly represents the world
(not just to some extent, but in all respects). Therefore, if we believe
of a family of theories that all are empirically adequate, but each
goes beyond the phenomena, then we are still free to believe that
each is false, and hence their common part is false. For that common
part is phraseable as: one of the models of one of those theories
correctly represents the world.

The theory which Leibniz* holds of the world, *TNE*, can nevertheless be stated, and I have already done so. Its single axiom can
be the assertion that *TN(0)* is empirically adequate: that *TN(0)*
has a model containing motions isomorphic to all the appearances. Since *TN(0)* can be stated in English, this completes the
job.

It may be objected that so stated, *TNE* does not look like a
physical theory. Indeed it looks metalinguistic. This is a poor objection. The theory is clearly stated in English, and that suffices.
Whether or not it is axiomatizable in some more restricted
vocabulary may be a question of logical interest, but philosophically
it is irrelevant. Secondly, if the set of models of *TN(0)* can be
described without metalinguistic resources, then the above statement
of *TNE* is easily turned into a non-metalinguistic statement too.
Not that this matters. The only important point here is that the
empirical import of a family of empirically equivalent theories is
not usually their common part, but can be characterized directly in
the same terms in which empirical adequacy is claimed.

§4. *Theories and their Extensions*

The objection may be raised that theories can seem empirically
equivalent only as long as we do not consider their possible extensions. When we consider their application beyond the originally
intended domain of application, or their combination with other

theories or hypotheses, we find that distinct theories do have different empirical import after all.[6] An imperfect example is furnished by Brownian motion, which established the superiority of the kinetic theory over phenomenological thermodynamics. This example is imperfect, for it was known that the two theories disagreed even on macroscopic phenomena over sufficiently long periods of time. Until the discovery of Brownian motion, it was thought that experiments would not yield 'fine' enough data to shorten sufficiently the period of time required to show the divergence of the theories.

A perfect example can be constructed as a piece of quite realistic science fiction: let us imagine that such experiments as Michelson and Morley's, which led to the rise of the theory of relativity, did not have their spectacular, actual, null-outcome, and that Maxwell's theory of electromagnetism was successfully combined with classical mechanics. In retrospect we realize that such a development would have upset even Newton's deepest convictions about the relativity of motion; but we can imagine it.

Electrified and magnetic bodies appear to set each other in motion although they are some distance apart. Early in the nineteenth century mathematical theories were developed treating these phenomena in analogy with gravitation, as cases of action at a distance, by means of forces which such bodies exert on each other. But the analogy could not be perfect: it was found necessary to postulate that the force between two charged particles depends on their velocity as well as on the distance.

Adapting the idea of a universal medium of the propagation of light and heat (the luminiferous medium, or ether) found elsewhere in physics, Maxwell developed his theory of the electromagnetic field, which pervades the whole of space:

> It appears therefore that certain phenomena in electricity and magnetism lead to the same conclusions as those of optics, namely, that there is an ethereal medium pervading all bodies, and modified only in degree by their presence ...[7]

The force on an electrified body is a force 'exerted by' this medium, and depends on the body's position and on its velocity. Maxwell's Equations describe how this field develops in time.

The difficulties with Maxwell's theory concerned the mechanics of this medium; and his ideas about what this medium was like, were not successful. But this did not blind the nineteenth century to the power and adequacy of his equations describing the electromagnetic

field. The consensus is perhaps expressed by Hertz's famous statement 'Maxwell's Theory is Maxwell's Equations'. It would therefore
not be appropriate to call Maxwell's theory a mechanical theory,
but it did have mechanical models. The existence of such models
follows from a mathematical result due to Koenig, as Poincaré
detailed in the preface of his *Electricité et Optique* and Chapter XII
of his *Science and Hypothesis*. There was this strange new feature,
however; the forces depend on the velocities, and not merely on the
accelerations. There was accordingly a spate of thought-experiments
designed to measure absolute velocity. The very simplest was
described by Poincaré:

Consider two electrified bodies; though they seem to us at rest, they are both
carried along by the motion of the earth; an electric charge in motion,
Rowland has taught us, is equivalent to a current; these two charged bodies
are, therefore, equivalent to two parallel currents of the same sense and these
two currents should attract each other. In measuring this attraction, we shall
measure the velocity of the earth; not its velocity in relation to the sun or the
fixed stars, but its absolute velocity.[8]

The frustratingly uniform null-outcome of all such experiments led
to the demise of classical physics and the advent of relativity theory.
But let us imagine that the classical expectations were not dis-
appointed. Imagine that values are found for absolute velocities;
specifically for the centre of gravity of the solar system. In that
case, it would seem, one of the theories $TN(v)$ may be confirmed
and all the others falsified. Hence those theories were not
empirically equivalent after all.

But the reasoning is spurious. The definition of empirical equivalence did not rely on the *assumption* that only absolute acceleration
can have discernible effects. Newton made the distinction between
sensible measures and apparent motions on the one hand, and true
motions on the other, without presupposing more than that basic
mechanics within which there are models for Maxwell's equations.
The assertion was that each motion in a model of $TN(v)$ is
isomorphic to a motion in a model of $TN(v + w)$, for all constant
velocities v and w. This assertion was the reason for the claim of
empirical equivalence. The question before us is whether that
assertion was controverted by those nineteenth-century reflections.

The answer is definitely *no*. The thought-experiment, we may
imagine, confirmed the theory that adds to TN the hypotheses:

$H0$ The centre of gravity of the solar system is at absolute rest.

E0   Two electrified bodies moving in parallel, with absolute velocity v, attract each other with force $F(v)$.

This theory has a consequence strictly about appearances:

CON   Two electrified bodies moving with velocity v relative to the centre of gravity of the solar system, attract each other with force $F(v)$.

However, that same consequence can be had by adding to *TN* the two alternative hypotheses:

Hw   The centre of gravity of the solar system has absolute velocity w.

Ew   Two electrified bodies moving with absolute velocity $v+w$ attract each other with force $F(v)$.

More generally, for each theory $TN(v)$ there is an electromagnetic theory $E(v)$ such that $E(0)$ is Maxwell's and all the combined theories $TN(v)$ plus $E(v)$ are empirically equivalent with each other.

There is no originality in this observation, of which Poincaré discusses the equivalent immediately after the passage I cited above. Only familiar examples, but rightly stated, are needed it seems, to show the feasibility of concepts of empirical adequacy and equivalence. In the remainder of this chapter I shall try to generalize these considerations, while showing that the attempts to explicate those concepts *syntactically* had to reduce them to absurdity.

§5. *Extensions: Victory and Qualified Defeat*

The idea that theories may have hidden virtues by allowing successful extensions to new kinds of phenomena, is too pretty to be left. Developed independently of the example in the last section it might yet trivialize empirical equivalence. Nor is it a very new idea. In the first lecture of his *Course de philosophie positive*, Comte referred to Fourier's theory of heat as showing the emptiness of the debate between partisans of calorific matter and kinetic theory. The illustrations of empirical equivalence have that regrettable tendency to date; calorifics lost. Federico Enriques seemed to place his finger on the exact reason when he wrote: 'The hypotheses which are indifferent in the limited sphere of the actual theories acquire significance from the point of view of their possible *extension*.'[9] And this suggests that after all, distinct theories can never be *really* empirically equivalent, because they may differ significantly in their extensions.

To evaluate this suggestion we must ask what exactly is an extension of a theory. Let us suppose as in the last section that experiments did indicate the combined theory $TN(0)$ plus $E(0)$. In that case we would surely say that mechanics had been successfully extended to electromagnetism. What, then, is a successful extension?

There were mechanical models of electromagnetic phenomena; and also of the phenomena more traditionally subject to mechanics. What we have supposed is that all these appearances could *jointly* find a home among the motions in a single model of $TN(0)$. Certainly, we have here an extension of $TN(0)$, but first and foremost we have a *victory*. We have an extension, for the class of models that may represent the phenomena has been narrowed to those which satisfy the equations of electromagnetism. But it is a victory for $TN(0)$ because it simply bears out the claim that $TN(0)$ is empirically adequate: all appearances can be identified with motions in one of its models.

Such victorious extensions can never distinguish between empirically equivalent theories in the sense in which that relation was described above, for such theories have exactly the same resources for modelling appearances. It follows logically from the italicized description in Section 2 that if one theory enjoys such a victory, then so will all those empirically equivalent to it.

So if Enriques's idea is to be correct at all, there must be other sorts of extensions, which are not victories. Let us suppose that a theory is confronted with new phenomena, and these are not even piece-wise identifiable with motions in the models of that theory. Must that old theory then suffer an unqualified defeat, and hope for nothing more than a survival as 'correct for a limited range', as approximating some fragment of some victorious new theory? There seems to be one possibility intermediate between victory and total defeat. The class of substructures called *motions* might, for example, be widened to a larger class; let us say, *pseudo-motions*. And the theory might be weakened, so that it would claim only that every appearance can be identified with a pseudo-motion.

This would be a defeat, for the claim that the old theory is empirically adequate has been rescinded. But still it may be called an extension rather than a replacement, for the class of models (the over-all structures within which motions and pseudo-motions are defined) has no new members added. It is therefore an extension, which is not a victory but anyway a qualified defeat.

It is not so easy to find an example of this kind of extension within the sphere of mechanics, but the following may be one. Brian Ellis constructed a theory in which no forces are postulated, but the available motions are the same as in Newton's mechanics plus the postulate of universal gravitation.[10] The effect of gravitational attraction is cunningly woven into the basic equations of motion of Ellis's theory. But Ellis has pointed out that Newton's theory has a certain kind of superiority in that, if the effect of gravitation is just slightly different, then Newton's theory is much more easily amended than his. In other words, if Newton's theory turned out wrong in its astronomical predictions, there is an obvious way to try and repair it, without touching his basic laws of motion.

It is possible to construe this as follows: the two theories are empirically equivalent, but Newton's allows of certain obvious extensions of the second sort. To see it this way, one has to take the law G of universal gravitation as defining the *motions* (described in terms of relative distances) in Newton's models: a motion is a set of trajectories for which masses and forces can be found such that Newton's laws of motion and G are satisfied. Then if evidence accrued in favour of an alternative postulate G' about gravity, the extension could proceed on the idea that the gravitational force is itself a function of some other factor, and by defining *pseudo-motions* as trajectories satisfying the suitably generalized law.

It will, however, be clear that the second sort of extension is a defeat. There is a certain kind of superiority perhaps in the ability to sustain qualified rather than total defeat. But it is a pragmatic superiority. It cannot serve to upset the conclusion that two theories are empirically equivalent, for it does not show that they differ in any way (not even conditionally, not even counterfactually) in their empirical import.

Let me close this section with an example of another sort of pragmatic superiority, which strikes me as quite similar.

Suppose that two proposed theories have different axioms, but turn out to have the same theorems (and the same models, and the same specification of empirical substructures). I do not suppose that anyone would think that these two theories say different things. Even so, there may be a recognizable superiority, which appears when we attempt to generalize them. An interesting example of this is given by Belinfante in his discussion of von Neumann's 'proof' that there can be no hidden variables in quantum-mechanical phenomena.[11] The observable quantities are represented by operators A, B, ... each of which has associated with it an infinite matrix $(A)_{ij}$ and also a function $\langle A \rangle$ which gives its expectation value $\langle A \rangle_\varphi$ in any state φ.

When he wrote his own theory, von Neumann could have chosen either of the following principles concerning combination of observable quantities to serve as an axiom:

1. $\langle aA + bB \rangle_\varphi = a\langle A \rangle_\varphi + b\langle B \rangle_\varphi$
2. $(aA + bB)_{ij} = a(A)_{ij} + b(B)_{ij}$

With a suitable choice of other axioms and definitions, the one not chosen as an axiom would have been derivable as a theorem. In fact, von Neumann chose 1. When he then came to the question of hidden variables, he showed that their existence would contradict the generalization of his basic axioms to states supplemented with hidden variables. However, it can easily be shown that any reasonable hidden variable theory must reject the generalization of 1, although it can accept 2. Had von Neumann chosen his axioms differently, he might well have reached the conclusion that 1 can be demonstrated for all quantum-mechanical states, but does not hold for the postulable underlying microstates—and hence, that there could be hidden variables after all.

Such pragmatic superiorities of one theory over another are of course very important for the progress of science. But since they can appear even between different formulations of the same theory, and also may only show up in actual defeat, they are no reflection on what the theory itself says about what is observable.

§6. *Failure of the Syntactic Approach*

Specific examples of empirical adequacy and equivalence should suffice to establish the correctness and non-triviality of these concepts; but we need an account of them in general. It is here that the syntactic approach has most conspicuously been tried, and has most conspicuously failed.

The syntactic explication of these concepts is familiar for it is the backbone of the account of science developed by the logical positivists. A theory is to be conceived as what logicians call a deductive theory, hence, a set of sentences (the theorems), in a specified language. The vocabulary is divided into two classes, the observational terms and the theoretical terms. Let us call the observational sub-vocabulary E. The empirical import of a theory T is identified as its set of testable, or observational, consequences; the set of sentences T/E, which are the theorems of T expressed in sub-vocabulary E. Theories T and T' are declared empirically equivalent exactly if T/E is the same as T'/E. An extension of a theory is just an axiomatic extension.

Obvious questions were raised and settled. A theory would seem not to be usable by scientists if it is not axiomatizable. Is T/E axiomatizable if T is? William Craig showed that, if the sub-vocabulary E is suitably specified, and T is recursively axiomatizable in its total vocabulary, then so is T/E in the vocabulary E.[12] Note that the question is interesting to logicians only if it is construed as being about axiomatizability in a *restricted* vocabulary. Of course, if T is axiomatizable and E suitably specifiable in English, then T/E is too. But logicians attached importance to questions about restricted vocabularies, and that was seemingly enough for philosophers to think them important too.

A more philosophical problem was apparently posed by the very distinction between observational and theoretical terms. Certainly in some way every scientific term is more or less directly linked with observation. When the distinction began to seem untenable, those who wished still to work with the syntactic scheme began to divide the vocabulary into 'old' and 'new' (or 'newly introduced') terms.[13]

But all this is mistaken. The empirical import of a theory cannot be isolated in this syntactical fashion, by drawing a distinction among theorems in terms of vocabulary. If that could be done, T/E would say exactly what T says about what is observable and what it is like, and nothing more. But any unobservable entity will differ from the observable ones in the way it systematically lacks observable characteristics. As long as we do not abjure negation, therefore, we shall be able to state in the observational vocabulary (however conceived) that there are unobservable entities, and, to some extent, what they are like. The quantum theory, Copenhagen version, implies that there are things which sometimes have a position in space, and sometimes have not. This consequence I have just stated without using a single theoretical term. Newton's theory implies that there is something (to wit, Absolute Space) which neither has a position nor occupies a volume. Such consequences are by no stretch of the imagination about what there is in the observable world, nor about what any observable thing is like. The reduced theory T/E is not a description of part of the world described by T; rather, T/E is, in a hobbled and hamstrung fashion, the description by T of everything.

Thus on the syntactic approach, the distinction between truth and empirical adequacy reduces to triviality or absurdity, it is hard to say which. Similarly for empirical equivalence. Recalling Section 2, we see that $TN(0)$ and TNE must be empirically equivalent, for the latter stated that $TN(0)$ is empirically adequate. But the former states that there is something (to wit, Absolute Space) which is different from every appearance by lacking even those minimal characteristics which all appearances share. Hence, $TN(0)/E$ is not the same as TNE/E; and so, on the syntactic approach, these theories are not empirically equivalent after all.

Philosophers seem to have been bothered more by ways in which the syntactic definition of empirical equivalence might be too broad. It was noted that many theories T are such that T/E is tautological, or wellnigh so. Such theories presumably derive their empirical import from the consequences they have when conjoined with other theories or empirical hypotheses. But in that case, T/E and T'/E might be the same even if T and T' are about totally different subjects.

To eliminate this embarrassment, extensions of theories were considered.[14] With a bow to Enriques, one may newly stipulate T and T' are empirically equivalent if and only if all their axiomatic extensions are, that is, if for every theory T'', (T plus T'')$/E$ is the same as (T' plus T'')$/E$.

While this manœuvre removes the second embarrassment, it runs afoul of the first. $TN(0)$ and TNE are again declared non-equivalent. Worse yet. $TN(0)$ is no longer empirically equivalent to the other theories $TN(v)$. This is shown by the examples of spurious reasoning in Section 4 above: $TN(0)$ plus $E(0)$ is not equivalent to $TN(v)$ plus $E(0)$ for non-zero values of v. But all the theories $TN(v)$ *are* empirically equivalent. Nor is it easy to see how we could restrict the class of axiomatic extensions to be considered so as to repair this deficiency.

These criticisms should suffice to show that the flaws in the linguistic explication of the empirical import of a theory are not minor or superficial. They do not, of course, constitute an *a priori* proof of the impossibility of a pure observation language. But such a project loses all interest when it appears so clearly that, even if such a language could exist, it would not help us to separate out the information which a theory gives us about what is observable. It seems in addition highly unlikely that such a language could exist. For at the very least, if it existed it might not be translatable into natural language. An observation language would be theoretically neutral at all levels. So if A and B are two of its simplest sentences, they would be logically independent. This shows at once that they could not have the English translations 'there is red-here-now' and 'there is green-here-now', which are mutually incompatible. Pursuing such questions further does not seem likely to shed any light on the nature or structure of science.

The syntactically defined relationships are simply the wrong ones. Perhaps the worst consequence of the syntactic approach was the way it focused attention on philosophically irrelevant technical questions. It is hard not to conclude that those discussions of axiomatizability in restricted vocabularies, 'theoretical terms', Craig's theorem, 'reduction sentences', 'empirical languages', Ramsey and Carnap sentences, were one and all off the mark—solutions to purely self-generated problems, and philosophically irrelevant. The main lesson of twentieth-century philosophy of science may well be this: no concept which is essentially language-dependent has any philosophical importance at all.

§7. *The Hermeneutic Circle*

We have seen that we cannot interpret science, and isolate its empirical content, by saying that our language is divided into two parts. Nor should that conclusion surprise us. The phenomena are saved when they are exhibited as fragments of a larger unity. For that very reason it would be strange if scientific theories described the phenomena, the observable part, in different terms from the rest of the world they describe. And so an attempt to draw the conceptual line between phenomena and the trans-phenomenal by means of a distinction of vocabulary, must always have looked too simple to be good.

Not all philosophers who have discussed the observable/unobservable distinction, by any means, have done so in terms of vocabulary. But there has been a further assumption common also to critics of that distinction: that the distinction is a philosophical one. To draw

it, they seem to assume, is in principle anyway the task of the philosophy of perception. To draw it, in principle anyway, philosophy
must mobilize theories of sensing and perceiving, sense data and
experiences, *Erlebnisse* and *Protokolsaetze*. If the distinction is a
philosophical one, then it is to be drawn, if at all, by philosophical
analysis, and to be attacked, if at all, by philosophical arguments.

This attitude needs a Grand Reversal. If there are limits to
observation, these are a subject for empirical science, and not for
philosophical analysis. Nor can the limits be described once and for
all, just as measurement cannot be described once and for all. What
goes on in a measurement process is differently described by classical
physics and by quantum theory. To find the limits of what is observable in the world described by theory T we must inquire into T itself,
and the theories used as auxiliaries in the testing and application
of T.

We have now come to the 'hermeneutic circle' in the interpretation
of science. I want to spell this out in detail, because one might too
easily get a feeling of vicious circularity. And I want to give specific
details on how science exhibits clear limits on observability.

Recall the main difference between the realist and anti-realist
pictures of scientific activity. When a scientist advances a new
theory, the realist sees him as asserting the (truth of the) postulates.
But the anti-realist sees him as displaying this theory, holding it up
to view, as it were, and claiming certain virtues for it.

This theory draws a picture of the world. But science itself
designates certain areas in this picture as observable. The scientist,
in accepting the theory, is asserting the picture to be accurate in
those areas. This is, according to the anti-realist, the only virtue
claimed which concerns the relation of theory to world alone. Any
other virtues to be claimed will either concern the internal structure
of the theory (such as logical consistency) or be pragmatic, that is,
relate specifically to human concerns.

To accept the theory involves no more belief, therefore, than that
what it says about observable phenomena is correct. To delineate
what is observable, however, we must look to science—and possibly
to that same theory—for that is also an empirical question. This
might produce a vicious circle if what is observable were itself not
simply a fact disclosed by theory, but rather theory-relative or
theory-dependent. It will already be quite clear that I deny this; I
regard what is observable as a theory-independent question. It is a

function of facts about us *qua* organisms in the world, and these facts may include facts about the psychological states that involve contemplation of theories—but there is not the sort of theory-dependence or relativity that could cause a logical catastrophe here.

Let us consider two concrete examples which have been found puzzling. The first, already mentioned by Grover Maxwell, concerns molecules. Certain crystals, modern science tells us, are single molecules; these crystals are large enough to be seen—so, some molecules are observable. The second was mentioned to me by David Lewis: astronauts reported seeing flashes, and NASA scientists came to the conclusion that what they saw were high-energy electrons.

Is there anything puzzling about these examples? Only to those who think there is an intimate link between theoretical terms and unobservable entities or events. Compare the examples with Eddington's famous table: that table is an aggregate of interacting electrons, protons, and neutrons, he said; but that table is easily seen. If a crystal or table is classified by a theory as a theoretically described entity, does the presence of this observable object become evidence for the reality of other, different but similarly classified entities? Everything in the world has a proper classification within the conceptual framework of modern science. And it is this conceptual framework which we bring to bear when we describe any event, including an observation. This does not obliterate the distinction between what is observable and what is not—for that is an empirical distinction—and it does *not* mean that a theory could not be right about the observable without being right about everything.

We should also note here the intertranslatability of statements about objects, events, and quantities. There is a molecule in this place; the event of there-being-a-molecule occurs in this place (this is, roughly, Reichenbach's event language); a certain quantity, which takes value *one* if there is a molecule here and value *zero* if there is not, has value *one*. There is little difference between saying that a human being is a good detector of molecules and saying that he is a good detector of the presence of molecules. Any such classification of what happens may be correct, relative to a given, accepted theory. If we follow the principles of the general theory of measurement used in discussions of the foundations of quantum mechanics, we call system Y a measurement apparatus for quantity A exactly if Y has a certain possible state (the ground-state) such that if Y is in that state and coupled with another system X in *any* of its possible states, the evolution of the combined system (X *plus* Y) is subject to a law of interaction which has the effect of correlating the values of A in X with distinct values of a certain quantity B (often called the 'pointer reading observable') in system Y. Since observation is a special subspecies of measurement, this is a good picture to keep in mind as a partial guide.

Science presents a picture of the world which is much richer in content than what the unaided eye discerns. But science itself teaches us also that it is richer than the unaided eye *can* discern. For science itself delineates, at least to some extent, the observable parts of the world it describes. Measurement interactions are a special subclass of physical interactions in general. The structures definable from measurement data are a subclass of the physical structures described. It is in this way that science itself distinguishes the observable which it postulates from the whole it postulates. The distinction, being in part a function of the limits science discloses on human observation, is an anthropocentric one. But since science places human observers among the physical systems it means to describe, it also gives itself the task of describing anthropocentric distinctions. It is in this way that even the scientific realist must observe a distinction between the phenomena and the trans-phenomenal in the scientific world-picture.

§8. *Limits to Empirical Description*

Are there limits to observation? While the arguments of Grover Maxwell aim to establish that in principle there are not (so as to undercut the very possibility of the statement of an empiricist philosophy of science), other arguments aim to establish the inadequacy of empiricism because of these limits. Since physical theory cannot be translated, without remainder, into a body of statements that describe only what the observable phenomena are like, such arguments run, empiricism cannot do justice to science. I grant the premiss, of course, and wish here to reinforce it by giving a more precise statement of the limits of empirical description, and some examples.

Before attempting precision, let us examine the standard example of 'underdetermination' to be drawn from foundational studies in classical mechanics. In the context of that theory, and arguably in all of classical physics, all measurements are reducible to series of measurements of time and position. Hence let us designate as basic observables all quantities which are functions of time and position alone. These include velocity and acceleration, relative distances and angles of separation—all the quantities used, for example, in reporting the data astronomy provides for celestial mechanics. They do not include mass, force, momentum, kinetic energy.

To some extent, and in many cases, these other quantities can be calculated from the basic observables. Hence the many proposed 'definitions' of force and mass in the nineteenth century, and the availability of axiomatic theories of mechanics today in which mass is not a primitive quantity.[15] But, as Patrick Suppes has emphasized, if we postulate with Newton that every body has a mass, then mass is not definable in terms of the basic observables (not even if we add force).[16] For, consider, as simplest example, a (model of mechanics in which a) given particle has constant velocity throughout its existence. We deduce, within the theory, that the total force on it equals zero throughout. But every value for its mass is compatible with this information.

What, then, of those 'definitions' of mass? The core of truth behind them is that mass is experimentally accessible, that is, there are situations in which the data about the basic observables, plus hypotheses about forces and Newton's laws, allow us to calculate the mass. We have here a *counterfactual*: if two bodies have different masses and if they *were* brought near a third body in turn, they *would* exhibit different acceleration. But as the example shows, there are models of mechanics—that is, worlds allowed as possible by this theory—in which a complete specification of the basic observable quantities does not suffice to determine the values of all the other quantities. Thus the same observable phenomena equally fit more than one distinct model of the theory. (Remember that empirical adequacy concerns actual phenomena: what does happen, and not, what would happen under different circumstances.)

I mentioned briefly the axiomatic theories of mechanics developed in this century. We see in them many different treatments of mass. In the theory of McKinsey, Sugar, and Suppes, as I think in Newton's own, each body has a mass. But in Hermes's theory, the mass ratio is so defined that if a given body never collides with another one, there is no number which is the ratio of its mass to that of any other given body. In Simon's, if a body X is never accelerated, the term 'the mass of X' is not defined. In Mackey's any two bodies which are never accelerated, are arbitrarily assigned the same mass.[17]

What explains this divergence, and the conviction of the authors that they have axiomatized classical mechanics? Well, the theories they developed are demonstrably empirically equivalent in exactly the sense I have given that term. Therefore, from the point of view of empirical adequacy, they are indeed equal. The thesis of constructive empiricism, that what matters in science is empirical adequacy, and not questions of truth going beyond that, explains this chapter in foundational studies.

In quantum mechanics we can find a similarly simple, telling example. First, I must make some preliminary remarks. The states are represented by vectors in a Hilbert space, and simple mathematical operations can be performed on these vectors. To calculate the probability of a measurement outcome, the theory tells us to proceed as follows. First we represent the state of the system by means of such a vector in a Hilbert space. Then we multiply that vector by a positive scalar, so that the result is a new vector just like the first, except that it has unit length. Next, we express this unit vector ψ in terms of a family of vectors (eigen-vectors) specially associated with the physical magnitude we are measuring, in this form:

$$\psi = c_1\psi_1 + \ldots + c_i\psi_i + \ldots$$

Each vector ψ_i corresponds to one possible measurement result r_i. The probability that the result will be r_k equals the square of coefficient c_k (or what corresponds to the square for complex numbers, if that coefficient is complex).

In view of this, it is often said that all positive multiples of ψ represent the same state. For if you begin with $k\psi$ or with $m\psi$, your first step will be to 'normalize', that is, multiply by a scalar so as to arrive at unit vector ψ, There is 'no physical difference', 'the phase has no physical significance' people say; and the reason they give is that the probabilities for measurement outcomes are the same.

Now consider a simple physical operation, rotation. Rotating a system changes its state. There are corresponding operations on vectors, to change the vector that represented the system before, to the one that represents it after the rotation. Let us call the vector operation that corresponds to a rotation through angle x by the name R_x. If the old state was ψ, the new state, after this rotation, is $R_x\psi$. In general, the probabilities for measurement outcomes are very different in this new state, so there is here in general a genuine physical difference.

One special case is the rotation through 2π radians, a complete circle. Physically, it brings the system back to its original position, in the classical, macroscopic examples we know so well. In the quantum analogue, the operation $R_{2\pi}$ is also quite simple: multiplication by the scalar -1. Hence $R_{2\pi}\psi = -\psi$. If we now expand this new vector in terms of the eigen-vectors ψ_i, we get the coefficients $-c_i$. But if we then calculate the probabilities of measurement results, we square these, so the minuses disappear. Those probabilities are therefore exactly the same for the new state as for the old.

Following the same reasoning as before, we should now say that $R_{2\pi}$, like positive scalar multiplication, simply produces a vector representing the same physical state as the original vector. But there has been a good deal of discussion of this case in the literature, and that apparently easy way out is not available to us.[18] To explain this, we must look at another operation on vectors, namely superposition. If φ and ψ are two vectors, then $(k\varphi + m\psi)$ is a superposition of them, which is again a vector in the same space, and also represents a physical state. We can sum up the argument in the literature as follows: if ψ and $R_{2\pi}\psi$ really represented exactly the same physical state, then the superposition $(k\varphi + m\psi)$ would represent the same state as $(k\varphi + mR_{2\pi}\psi)$. That the latter is not so, is easily seen by calculating probabilities for various observables. Klein and Opat designed an experiment on a beam of neutrons in which the observable differences between the two sorts of superpositions were verified: a Fresnel diffraction experiment in which the diffracting object was the boundary between two regions carrying opposite magnetic fields.

What should we conclude from this? The case is quite similar to that of classical mass. If in one possible world, an isolated system is in state ψ and in another it is in state $R_{2\pi}\psi$, no amount of empirical information actually available can tell the observer which of these two worlds he is in. But there is a *counterfactual* statement we are inclined to make about the case: if the system had interacted with another one in such and such a way, the results would have been different in the two cases. The observable phenomena which are actual, however, are the same.

The literature on the measurement problem in quantum mechanics contains a great deal more tantalizing discussion of the extent to which the observable macroscopic phenomena 'underdetermine' the underlying microscopic state. I refer specifically to Nancy Cartwright's conclusion, based on the quantum thermodynamic approach of Daneri, Loinger, and Prosperi, that a certain superposition of states is indistinguishable in measurement with respect to all macroscopic observables, from a corresponding mixture.[19] Again it is impossible to say, really there is no underdetermination because the two are states between which there is no physical difference. For if systems in these two states *were* subject to interaction with a special third sort of system, the results *would be* different. (This is analogous to the similar point about the masses of actually unaccelerated bodies—the physical difference comes out in the counterfactual assertions we base on what the theory says about what would happen in other, non-actual conditions.) But I am here touching on large and complex issues, and it is hard to say something which is at once simple and uncontroversial.

For the theory of general relativity, we have two studies by Clark Glymour that clearly bring out limits of observation. The first assumes reasonably that measurement divulges only the values of local quantities, and then shows that measurement cannot uniquely determine the global structure of space-time.[20] The second arrives at the same conclusion from the assumption that any observed structure must lie in the absolute past cone of some space-time point.[21] But it is the theory of relativity itself, surely, that forces these assumptions on us, for it forces us to locate observers in space-time, and restricts the information that can reach them.

In this section I have tried to give examples of a very basic and general sort of how, in the description of the world by a physical theory, we can see a division between that description taken as a whole, and the part that pertains to what is observationally determined. The limitations exhibited reach very deeply into the theories in question, and do not relate merely to such 'accidental' limitations as perceptual thresholds and humanly available energy. Realists are generally a bit ambiguous in their feelings toward such limitations. On the one hand they want to emphasize them and say that as a consequence, there is much more to the world described by physics than is dreamt of in the empiricist's philosophy. On the other, they wish to play down the underdetermination, arguing that any precise definition of empirical adequacy and empirical equivalence will lead to the conclusion that a physical theory is completely adequate only if it is true. My view is that physical theories do indeed describe much more than what is observable, but that what matters is empirical adequacy, and not the truth or falsity of how they go beyond the observable phenomena. And the precise definition of empirical adequacy, because it relates the theory to the *actual* phenomena (and not to anything which *would* happen if the world *were* different, assertions about which have, to my mind, no basis in fact but reflect only the background theories with which we operate) does not collapse into the notion of truth.

§9. *A New Picture of Theories*

Impressed by the achievements of logic and foundational studies in mathematics at the beginning of this century, philosophers began to think of scientific theories in a language-oriented way. To present a theory, you specified an exact language, some set of axioms, and a partial dictionary that related the theoretical jargon to the observed phenomena which are reported. Everyone knew that this was not a very faithful picture of how scientists do present theories, but held that it was a 'logical snapshot', idealized in just the way that point-masses and frictionless planes idealize mechanical phenomena. There is no doubt that this logical snapshot was very useful to philosophical discussion of science, that there was something to it, that it threw light on some central problems. But it also managed to mislead us.

A picture is only a picture—something to guide the imagination as we go along. I have proposed a new picture, still quite shallow, to guide the discussion of the most general features of scientific theories. To present a theory is to specify a family of structures, its *models*; and secondly, to specify certain parts of those models (the *empirical substructures*) as candidates for the direct representation of observable phenomena. The structures which can be described in experimental and measurement reports we can call *appearances*: the theory is empirically adequate if it has some model such that all appearances are isomorphic to empirical substructures of that model. I am certainly not the first to propose this picture: you can see it at work for example in the writings of Wojcicki and Przełewski in Poland, Dalla Chiara and Toraldo di Francia in Italy, Suppes and Suppe in America.[22] (For instance, what Patrick Suppes calls *empirical algebras* are instances of what I call *appearances*, and he relates them to parts of the models, and thus describes the relation of theory to data, in much the way I have outlined.)

The form in which theories are actually presented in the technical literature is of course not a sure guide to the form we should conceive them to have. Yet I would still claim support for this proposed way of looking at theories from the actual form of presentation, and indeed from those presentations of theories that are most likely to support the opposing view: the axiomatic ones. In many texts and treatises on quantum mechanics, for example, we find a set of propositions called the 'axioms of quantum theory'. They do not look very much like what a logician expects axioms to look like; on the contrary, they form, in my opinion, a fairly straightforward description of a family of models, plus an indication of what are to be taken as empirical substructures of these models:

Axiom I   To every pure state corresponds a vector, and to all the pure states of a system, a Hilbert space of vectors.

Axiom II   To every observable (physical magnitude) there corresponds a Hermitean operator on this Hilbert space.

Axiom III The possible values of an observable are the eigenvalues of its corresponding operator.

Axiom IV   The expectation value of observable A in state W equals the trace $Tr(AW)$.

To think that this theory is here presented axiomatically in the sense that Hilbert presented Euclidean geometry, or Peano arithmetic, in axiomatic form, seems to me simply a mistake.

Those axioms are instead a description of the models of the theory plus a specification of what the empirical substructures are. The ones I have given are only the beginning for quantum theory of course. For example, as next step there will be principles laid down that say which operator represents energy, or momentum, or how two operators representing two given observable quantities (such as position and momentum) are related to each other. In this further development there is no *a priori* right and wrong; the theory is successfully continued if we can find *some* Hermitean operator to represent energy; and so forth.

When Patrick Suppes first advocated this sort of picture of theories in his studies of mechanics (with the slogan that *philosophy of science should use mathematics, and not meta-mathematics*) he proposed a canonical form for the formulation of theories. This used set theory. To present classical mechanics, for instance, he would give the definition: 'A system of classical mechanics is a mathematical structure of the following sort ...' where the dots are replaced by a set-theoretic predicate. Although I do not wish to favour any mathematical presentation as the canonical one, I am clearly following here his general conception of how, say, the theory of classical mechanics is to be identified.

Looking at Suppes's formulation, it is easy to discuss two points that might otherwise be puzzling. How could, for example, classical mechanics have a model into which all phenomena can be embedded, when that theory does not even mention electricity? The answer is that a mathematical structure might be a system of this or that and also have much structure that does not enter the description of that sort of system. To be a system of mechanics, for example, it must have a set of entities plus a function which assigns each of those a velocity at each instant in time. Well, it could also have in it a function that assigns each of those entities an electric charge; it would still be a system of mechanics, as well as, say, a system of electrodynamics. The second question concerns unintended realizations. Could it not be that some system of mechanics happens to be also a system of optics if we re-label its constituents in a certain way? Well, not in that example perhaps, but there can be examples of that sort. The same formula may govern diffusion of gases and of heat. So perhaps a theory could fail to be empirically adequate as intended, but be so when the phenomena are embedded in its models in an unexpected way? Certainly that is possible.

This suggests that the intention, of which sorts of phenomena are to be embedded in what kinds of empirical substructures, be made part of the theory. I do not think that is necessary. Unintended realizations disappear when we look at larger observable parts of the world; say, the optics and mechanics of moving light sources together. If for a little while some fairly weak theory is empirically adequate, but in a way its advocates are not in a position to notice, that seems hardly an important or frequent enough occurrence to guard against by more complex definitions.

Let me also mention, to complete this part of the discussion, that while I consider Suppes's account of the structure of scientific theories an excellent vehicle for the elucidation of these general distinctions, I do regard it still as relatively shallow. In this book I am mainly concerned with the relation between physical theories and the world rather than with that other main topic, the structure of physical theory. With respect to the latter I see two main lines of approach: one deriving from Tarski and brought to maturity by Suppes and his collaborators (the *set-theoretic structure approach*) and the other initiated by Weyl and developed by Evert Beth (the *state-space approach*). The first is adamantly extensionalist, the second gives a central role to modality. Each was somewhat language-oriented to begin with, buth both shed these linguistic trappings as they were developed. My own inclination in that subject area has been toward the state-space approach. The general concepts used in the discussion of empirical adequacy, in this chapter, pertain to scientific theories conceived in either way.

Having insisted that the new picture of theories constitutes a radical break with the old, I wish to conclude by outlining some of its peculiar features. Of course, it too provides an idealization: only in foundational studies in physics do we see the family of models carefully described, and only when paradox threatens (as in the measurement problem for quantum mechanics) does anyone try to be very precise about the relation between theory and experiment. That is what it is to be healthy; philosophy is professionally morbid. Still, it is reasonable to draw distinctions and define theoretical relations in terms of the idealization.

If for every model M of T there is a model M' of T' such that all empirical substructures of M are isomorphic to empirical substructures of M', then T is *empirically at least as strong as* T'. Let us abbreviate this to: $T >_e T'$.

We may put this as follows: empirical adequacy, like truth, is 'preserved under watering-down'. I can water a theory down *logically* by disjoining it with some further hypothesis: thus Ptolemy watered down Aristotle's theory of the heavens by asserting that planets did certainly move along circles, but those circles need not have stationary centres. If A is true, so is (A or B). Similarly we can water down a theory *empirically*, either by admitting some new models, or by designating some new parts as empirical substructures in the old models, or both.

Logical strength is determined by the class of models (inversely: the fewer the models the (logically) stronger the theory!) and empirical strength is similarly determined by the classes of empirical substructures. If $T >_e T'$ and $T' >_e T$, then they are *empirically equivalent.* We may call a theory *empirically minimal* if it is empirically non-equivalent to all logically stronger theories—that is, exactly if we cannot keep its empirical strength the same while discarding some of the models of this theory.

The notions of empirical adequacy and empirical strength, added to those of truth and logical strength, constitute the basic concepts for the semantics of physical theories. Of course, this addition makes the semantics only one degree less shallow than the one we had before. The semantic analysis of physical theory needs to be elaborated further, preferably in response to specific, concrete problems in the foundations of the special sciences. Especially pressing is the need for more finely delineated concepts pertaining to probability for theories in which that is a basic item. I shall return to this subject in another chapter below.

Empirical minimality is emphatically *not* to be advocated as a virtue, it seems to me. The reasons for this point are pragmatic. Theories with some degree of sophistication always carry some 'metaphysical baggage'. Sophistication lies in the introduction of detours via theoretical variables to arrive at useful, adequate, manageable descriptions of the phenomena. The term 'metaphysical baggage' will, of course, not be used when the detour pays off; it is reserved for those detours which yield no practical gain. Even the useless metaphysical baggage may be intriguing, however, because of its potentialities for future use. An example may yet be offered by hidden variable theories in quantum mechanics.[23] The 'no hidden variables' proofs, as I have already mentioned, rest on various assumptions which may be denied. Mathematically speaking there exist hidden variable theories equivalent to orthodox quantum theory in the following sense: the algebra of observables, reduced *modulo* statistical equivalence, in a model of the one is isomorphic to that in a model of the other. It appears to be generally agreed that such theories confront the phenomena exactly by way of these algebras of statistical quantities. On that assumption, theories equivalent in this sense are therefore empirically equivalent. Such hidden variable models have much extra structure, now looked upon as 'metaphysical baggage', but capable of being mobilized should radically new phenomena come to light.

With this new picture of theories in mind, we can distinguish between two epistemic attitudes we can take up toward a theory. We can assert it to be true (i.e. to have a model which is a faithful replica, in all detail, of our world), and call for belief; or we can simply assert its empirical adequacy, calling for acceptance as such. In either case we stick our necks out: empirical adequacy goes far beyond what we can know at any given time. (All the results of measurement are not in; they will never all be in; and in any case, we won't measure everything that can be measured.) Nevertheless there is a difference: the assertion of empirical adequacy is a great deal weaker than the assertion of truth, and the restraint to acceptance delivers us from metaphysics.

4

Empiricism and Scientific Methodology

> I shall add only the fantasy that God or Nature may be
> playing thousands, perhaps a countless number, of simul-
> taneous Eleusis games with intelligences on planets in the
> universe ... Prophets and False Prophets come and go, and
> who knows when one round will end and another begin?
> Searching for any kind of truth is an exhilarating game.
> It is worth remembering that there would be no game at
> all unless the rules were hidden.
>
> Martin Gardner, 'On Playing New
Eleusis', Scientific American,
October 1977

So far I have concentrated on what a theory is and how it is to be understood. But from an empiricist point of view, the construction of theories cannot be the supreme scientific activity; at least not in the hierarchical sense that everything else is subordinate to it. For theories do much besides answer the factual questions about regularities in the observable phenomena which, according to empiricism, are the scientist's basic topic of concern. This is intelligible only if the other aspects of theorizing can be understood as instrumental for the pursuit of empirical strength and adequacy, or for the serving of other aims which are not basic but still part of the scientific enterprise.

In this chapter I shall address four main questions:

(1) does the rejection of realism presuppose or entail an epistemology that will lead to a self-defeating scepticism? (2) is the methodology of science and experimental design intelligible on any but a realist interpretation of science? (3) is the ideal of the unity of science, or even the practice of using distinct scientific theories in conjunction, intelligible on an empiricist view of science? and (4) what sense can we make of theoretical virtues (such as simplicity, co-

herence, explanatory power) which are not reducible to empirical adequacy or empirical strength?

The last question leads inevitably to the topic of scientific explanation, an account of which I shall present after this chapter. But for the other questions too, I shall not merely attempt a defence against possible realist objections, but hope to present a constructive empiricist alternative.

§1. *Empiricist Epistemology and Scepticism*

In Chapter 2 I objected to various lines of reasoning which would lead one to scientific realism. Some of these arguments, however, concerned the acceptance of a hypothesis or theory *as true*, on the basis of the evidence that bears it out. I resisted such inference, arguing in effect that when the theory has implications about what is not observable, the evidence does not warrant the conclusion that it is true.

The danger is clearly that, by parity of reasoning, my arguments would, if successful at all, establish that the evidence never warrants a conclusion that goes beyond it. This is already quite unacceptable, for we do in our daily life infer, or at least arrive at, conclusions that go beyond the evidence we have, and will resist as sophistical any philosophical theory which calls us irrational for that reason alone.

The danger is, I think, only apparent, but it does draw attention to important issues in epistemology that also divide realists and anti-realists. I would go too far afield if, at this point, I attempted to deal with these issues in anything like a thorough fashion.[1] But if I must postpone to another occasion a treatise on epistemology, I must at least defend myself against this threatened scepticism.

When I discussed the putative rule of inference to the best explanation, which must indeed be unacceptable to an anti-realist, I offered an alternative. The alternative is that explanatory power is certainly *one* criterion of theory choice. When we decide to choose among a range of hypotheses, or between proffered theories, we evaluate each for how well it explains the available evidence. I am not sure that this evaluation will always decide the matter, but it may be decisive, in which case we choose to accept that theory which is the best explanation. But, I add, the decision to accept is a decision to accept as empirically adequate. The new belief formed is not that the theory is true (nor that it gives a true picture of what there is and of what is going on plus approximately true numerical information), but that the theory is empirically adequate. In the case of a hypothesis, the belief formed is that the theory which results from the one we have already accepted, by adding this hypothesis, is empirically adequate.[2]

When the hypothesis is solely about what is observable, the two procedures amount to the same thing. For in that case, empirical adequacy coincides with truth. But clearly this procedure leads us to conclusions, about what the observable phenomena are like, which go beyond the evidence available. Any such evidence relates to what has already happened, for example, whereas the claim of empirical adequacy relates to the future as well.

At this point, it may be objected that I have drawn an arbitrary line. Surely the observable objects and processes we recognize in our world, are also postulated entities, believed in because they best explain and systematize the sense-experience or series of sense-data which are at bottom the only real evidence we have? Should I not be as unwilling to postulate tables and trees as forces, fields, and Absolute Space, unless I have a rationale that shows them to be essentially different in some relevant way?

I mention this objection because I have heard it, but it astonishes me since philosophers spent the first five decades of this century refuting the presuppositions that lie behind it. Indeed, every school of thought in Western philosophy, Continental as well as Anglo-Saxon, refuted them in its own terms. But it is easy for me to add at least this: such events as experiences, and such entities as sense-data, when they are not already understood in the framework of observable phenomena ordinarily recognized, are theoretical entities. They are, what is worse, the theoretical entities of an armchair psychology that cannot even rightfully claim to be scientific. I wish merely to be agnostic about the existence of the unobservable aspects of the world described by science—but sense-data, I am sure, do not exist.

There does remain the fact that even in endorsing a simple perceptual judgement, and certainly in accepting any theory as empirically adequate, I am sticking my neck out. There is no argument there for belief in the truth of the accepted theories, since it is not an epistemological principle that one might as well hang for a sheep as for a lamb. A complete epistemology must carefully investigate the conditions of rationality for acceptance of conclusions that go beyond one's evidence. What it cannot provide, I think (and to that extent

I am a sceptic), is rationally compelling forces upon these epistemic decisions.

However, there is also a positive argument for constructive empiricism—it makes better sense of science, and of scientific activity, than realism does and does so without inflationary metaphysics.

§2. *Methodology and Experimental Design*

§2.1 *The Roles of Theory*

The real importance of theory, to the working scientist, is that it is a factor in experimental design.

This is quite the reverse of the picture drawn by traditional philosophy of science. In that picture, everything is subordinate to the aim of knowing the structure of the world. The central activity is therefore the construction of theories that describe this structure. Experiments are then designed to test these theories, to see if they should be admitted to the office of truth-bearers, contributing to our world-picture.

Whatever the core of truth in that picture (and surely it has some truth to it) it contrasts sharply with the activity Kuhn has termed 'normal science', and even with much of what is revolutionary. Scientists aim to discover facts about the world—about the regularities in the observable part of the world. To discover these, one needs experimentation as opposed to reason and reflection. But those regularities are exceedingly subtle and complex, so experimental design is exceedingly difficult. Hence the need for the construction of theories, and for appeal to previously constructed theories to guide the experimental inquiry.

As Duhem already emphasized, the very search for new and deeper empirical regularities becomes couched in theoretical language. In the next subsection I shall describe the experiment by which Millikan measured the elementary electrical charge (of which all electrical charges are integral multiples). It will not be denied, I suppose, that his result answered many questions about regularities in the observable phenomena of electricity. Theory enters in two ways. The first is that the form his answer takes is that of a theoretical statement: *he is filling in the blanks in a developing theory*. The second is the role of already accepted theory in the design of his apparatus. This second role is the one I am at present emphasizing—it is the role that makes theory of value to the working scientist, as I said. The question was: 'What is the elementary electrical charge?' The reason the scientist turns to a theory to rely on is that he must first obtain an answer to the preceding question 'How can we experimentally determine the elementary electrical charge?'

If the above is correct, then the intimately intertwined development of theory and experimentation is intelligible from an empiricist point of view. For theory construction, experimentation has a twofold significance: testing for empirical adequacy of the theory as developed so far, *and* filling in the blanks, that is, guiding the continuation of the construction, or the completion, of the theory. Likewise, theory has a twofold role in experimentation: formulation of the questions to be answered in a systematic and compendious fashion, *and* as a guiding factor in the design of the experiments to answer those questions.[3] In all this we can cogently maintain that the aim is to obtain the empirical information conveyed by the assertion that a theory is or is not empirically adequate.

§2.2 *Measuring the Charge of the Electron*

Abstract descriptions of scientific aims and activity such as are given by realists and empiricists, may or may not sound plausible. If we go back to the things themselves, to use Husserl's phrase, we immediately revert to the naïve, unreflective descriptions of newspaper and magazine science writing. Herschel discovered Uranus, J. J. Thomson the electron, James Chadwick the neutron, Columbus America, and James Clerk Maxwell the electromagnetic field. The phenomenology of scientific theoretical advance may indeed be exactly like the phenomenology of exploration and discovery on the Dark Continent or in the South Seas, in certain respects. And it is also appropriate to talk in this fashion while immersed in the theoretical picture that guides the actual scientific work.

But let us step back for a moment and ask what roles experimentation and controlled observation play *vis-à-vis* the enterprise of constructing empirically adequate theories. One role philosophers have emphasized a great deal: the use of experimentation for theory testing. There are classic cases: Dominic Cassini's attempt to measure the curvature of the earth in order to adjudicate between Newtonian and Cartesian physics, Halley's prediction of the comet's return and its observation, the famous watch at the eclipse that bore out Einstein's theory implying the deflection of light rays in the gravitational field. This sort of experimental activity fits neatly into the empiricist's scheme, for clearly it is designed to test claims of empirical adequacy.

But this is not the sort where we use the terminology of discovery. There are cases where a theory says that there must be some entity or value, satisfying certain conditions, and experimental scientists *discover what that is*. Darwin's theory implied that there must be 'missing links', but could say very little about them. The search for the missing link, once followed so avidly by the general public, held many discoveries that were surprising but satisfyingly in accord with theory: the Java man and the Peking man, not to mention the Piltdown man.

Missing links in biological theory are observable entities of course, so we have in part a testing of claims of empirical adequacy. That history will not fit any of Darwin's models, unless such links existed. But the discoveries brought much new information which went into subsequent textbooks. When in physics we have a parallel discovery—the electron, the neutron, the magnitude of the charge of the electron—we obtain similarly new information that was not implied by the theory beforehand. This is, in part, information about what the unobservable entities are like—but surely, unless they exist, there can be no information about them to be had?

The answer to this objection consists in taking a purely functional view of what is happening. Atomic physics was developing slowly, as a theory, and at each stage, many blank spaces had to be left in the theory. Rather than fill such a blank with a conjectured answer, as hypothesis, and then testing the hypothesis, one carries out an experiment that shows *how the blank is to be filled if the theory is to be empirically adequate*. Then it is filled, and the theory construction has got one more step forward, and soon there are new consequences to be tested and *new blanks to be filled*. This is how experimentation guides the process of theory construction, while at the same time the part of the theory that has already been constructed guides the design of the experiments that will guide the continuation.

Between 1907 and 1911 Robert Millikan designed a new experimental approach to the measurement of the charge on the electron. While the success of his experiments in singling out a unique value for this charge is simultaneously a test of the theory that there exists this elementary electrical charge, it was not surprising at this time that such tests should bear out that theory. What made his experiments famous was the accuracy and unambiguousness in which they determined a unique value for this quantity whose value was theoretically an open question.

The main part of the apparatus was a circular chamber whose roof and floor consisted of two brass plates, 22 cm in diameter, and whose wall was a strip of thin sheet ebony, 16 mm high, containing three glass windows. A light beam could enter through one window and exit through another, while the third window was used for observation. Fine droplets of oil could enter the chamber from above, and due to the combined action of gravity and air resistance, would sink toward the lower plate with constant velocity. When the beam of light was switched on, such a droplet could be observed, with the appearance of a brilliant star on a black background.

The brass plates were connected to a battery and switch arrangement, which could be used to create an electrical field of strength between 3,000 and 8,000 volts per centimetre between the plates. (At this point, I use the terminology of previously well-established theory of electricity, on the macro-level, to describe what is happening. The check that this 'field is created' consists simply in reading a voltmeter.) Some droplets would, when the field was on, rise 'against gravity' toward the upper plate. By short-circuiting the plates just before the droplet strikes the roof, this effect disappears, and the droplet sinks again. Repeating this procedure, one such droplet was watched for a period of four and a half hours.

Background theory told us to expect some droplets to rise when the field was switched on, because some would naturally receive a charge due to friction. Theory also tells us that there may be variations in the speed of rising, and even that this speed will sometimes be zero (the droplet hovers in one spot). This is because, according to this background theory, the droplet may catch an ion from those which exist normally in the air. So far, the experiment bears out observational consequences of the theory—what is happening fits well into various models provided by the theory, since all these variations are observed in some droplets.

But now we can furthermore use the established part of the theory, and the observations on the speed of rising, to calculate the charges on the droplets. The apparent mass of the droplet is the difference between the actual mass and the buoyancy of the air; call this m. Let its charge at a given time be e, its speed under gravity v, and its speed when the electric field F is on w; the relations among these quantities are given by the equation

$$\frac{v}{w} = \frac{mg}{Fe - mg}$$

Since all but e are known, we can calculate e.

When a variation in rising speed occurs, this must then be attributed to a change in the charge, from e to e', say. If electrical charge comes only in multiples of a unit u, the charge on the electron, there must be a number k such that $e' - e = ku$. Gathering sufficient data of this sort, Millikan arrived at a mean value for u, which is very close to the one at present accepted.

Because of the way I have retold the story, what Millikan was doing now sounds exactly like what I said he was doing: that is, filling in a value for a quantity which, in the construction of the theory, was so far left open. Hence, in this case, *theory construction consists in experimentation*. And while it may be natural to use the terminology of discovery to report Millikan's results, the accurate way to describe it is that he was writing theory by means of his experimental apparatus. In a case such as this one, *experimentation is the continuation of theory construction by other means*. The appropriateness of the means follows from the fact that the aim is empirical adequacy.

§2.3 *Boyd on the Philosophical Explanation of Methodology*

In Section 2.1 above, I attributed two characters to the interaction between theory and experiment. On the one hand, theory is a factor in experimental design; on the other, experimentation is a factor in theory construction. The experiment by Millikan just discussed illustrates the *second*, and I gave an empiricist account of what goes on in such a case.

Richard Boyd has emphasized the *first* character of that interaction, that is, the role of theory in experimental design. Along the way we saw ample illustration of this in Millikan's design of his experiments: the established theory concerning electric fields, forces on electrified particles, and ionization served to design his experiment. And Boyd maintains that only scientific realism can help us make sense of this aspect of scientific activity.[4]

Boyd advocates scientific realism as an explanation, indeed as the only reasonable explanation, 'for the legitimacy of ... "intertheoretic" considerations in scientific methodology' (p. 7). Accepted theories play a role in experimental design—this role, according to Boyd, can only be described within the scientific realist's account of scientific activity, implying assumptions (if only *pro tem.*) concerning the truth of the theories used. The guesses based on these assumptions 'are so good that they are central to the success of the experimental method. What explanation besides scientific realism is possible?' (p. 12).

His argument has two parts, neatly corresponding to principles he labels as (1) and (2). The first is a thesis he attributes to anti-realists and is at pains to refute:

(1) If two theories have exactly the same deductive observational con-sequences, then any experimental evidence for or against one of them is evidence of the same force for or against the other. (Ibid., p. 2.)

Not surprisingly, Boyd finds that (1) is trivially false as it stands (con-sider theories which have no observation predicates at all, or whose observation consequences are all tautological). So to be debated, (1) must be amended by having the relation of empirical equivalence characterized in some more satisfactory way. Boyd proposes three amendments, each of which is also absurdly unsatisfactory. He con-cludes that (1) is untenable.

None of the characterizations of empirical equivalence of theories which Boyd examines is at all like the one I have proposed in these pages. Instead he uses logical or syntactic characterizations of the sort which we found wanting in the positivist account. Whatever the merits of (1) if understood to mean 'any two empirically equivalent theories are equally supported or contravened by the evidence' in the sense I would give to those terms, none of Boyd's arguments against it are telling.

The second principle is one which Boyd takes all philosophers to accept:

(2) Suppose that some principle of scientific methodology contributes to the reliability of that methodology in the following minimal sense: that its operation contributes to the likelihood that the observational con-sequences of accepted scientific theories will be (at least approximately) true. Then it is the business of scientific epistemology to *explain* the reli-ability of that principle. (Ibid., p. 3.)

Well, if we are not to cavil at the vagueness of the sentiment expressed, I suppose we ought to subscribe to it.

This second principle is itself about principles of scientific meth-odology and Boyd has a specific example in mind:

(P) a proposed theory T must be experimentally tested under situations repre-sentative of those in which, in the light of collateral information, it is most likely that T will fail, if it's going to fail at all. (Ibid., p. 10.)

This is, as it stands, innocuous, but the difference of opinion comes in the discussion of what 'in the light of collateral information' means. Boyd construes it as 'in the light of available theoretical knowledge' (p. 11). I imagine that he is using 'knowledge' lightly; he is referring to the account of underlying causal mechanisms implied by the accepted theories which form the background to the experimentation.

In illustration he gives two related schematic examples. A theory L is proposed to the effect that, by some chemical mechanism M, antibiotic A dissolves the cell walls of bacteria of sort C. From L plus appropriate chemical and bacteriological (previously accepted) information, an equation is derived expressing the population of bacteria C in a certain environment as a function of their initial population, the dosage of A, and the time elapsed since exposure to A. What sorts of considerations must guide the design of experiments to establish the acceptability of L?

Example 1: A drug similar to A is known (Boyd's term) to affect those bacteria not by dissolving their cell walls but by interfering with the development of new cell walls after mitosis. This makes it imperative to check the implication of L that A works by dissolving cell walls and not in that alternative fashion. Hence the bacterial population should be observed during a time interval too short for the typical bacterial cell to divide, while subjected to a dosage of A large enough, according to L, to kill a large proportion of the bacteria in this interval (if there is such an interval and dosage of course).

Example 2: The bacteria in question are known to be particularly prone to mutations affecting cell wall structure. This raises the possibility that L will fail when the time interval is long enough, and dosage of A low enough, to allow the selective survival of mutations with cell walls resistant to M. Hence a corresponding experiment, different from the one in Example 1, is indicated.

For 'known' we must of course read 'implied by a previously accepted theory'. The point remains that those established theories point to ways in which L may be false (by raising alternatives to L which would also explain data gathered on medium dosage and medium time intervals, presumably). In this way they guide the design of the sort of experimental test which L must pass in order to qualify as acceptable.

The point is that among the criteria for the adequacy of the experimental testing of a theory is this one: that it should be inquired, in the light of available theoretical knowledge, under what circumstances the causal claims made by the theory might plausibly go wrong, either because alternative causal mechanisms . . . might be operating instead of those indicated by the theory, or because causal mechanisms of sorts already known might plausibly be expected to interfere with those required by the theory in ways which the theory does not anticipate. (Ibid., p. 11.)

But, I suggest, the only explanation for this principle lies in a realistic understanding of the relevant collateral theories. (Ibid., pp. 11–12.)

We must admit that this is one explanation: that the collateral theories are believed to be true. But Boyd needs to establish not only that, as a realist, he can explain what is happening, but also that competing explanations are not feasible.

Let us see then, on Boyd's behalf, how an empiricist can render this methodology intelligible. In the above examples, the collateral theories suggested ways in which the function governing population decrease, in terms of drug dosage and time elapsed, might prove to be observably false. Boyd's point is no doubt that the manner in which those theories suggested these consequences, was by suggesting alternative underlying mechanisms which are not directly observable.

I would put this as follows: the models of L are quite simple, and reflection on the models of the collateral theories suggests ways in which the models of L could be altered in various ways. The empirical adequacy of L requires that the phenomena (bacterial population size and its variation) can be fitted into some of its models. Certain phenomena do fit the suggested altered models and not the models of L as it stands. Thus a test is devised that will *favour L* (or not favour it) *as against one of those contemplated alternatives*. But it is easy to see that what such a test will do is to speak for (or against) the empirical adequacy of L in those respects in which it differs from those alternatives.

The talk of underlying causal mechanisms can be construed therefore as talk about the internal structure of the models. In contrast with the logical, syntactic construal of theories which Boyd used in the discussion of what he called principle 1, we must direct our attention to the family of models of the theory to make sense of the pursuit of empirical adequacy through total immersion (for practical purposes) in the theoretical world-picture.

§2.4 *Phenomenology of Scientific Activity*

The working scientist is totally immersed in the scientific world-picture. And not only he—to varying degrees, so are we all. If I call a certain box a VHF receiver, if I call a fork electro-plated, if I so much as decide to turn on the microwave oven to heat my sandwich in the cafeteria, I am immersed in a language which is thoroughly theory-infected, living in a world my ancestors of two centuries ago could not enter.

In the language-oriented philosophy of science developed by the logical positivists, one could not say this while remaining empiricist. For the empirical import of a theory was defined via a division of its (*sic*) language into a theoretical and a non-theoretical part. This division was a philosophical one, that is, imposed from outside. And you could not limit your endorsement to the empirical import of the theory unless your language remained in principle limited to the non-theoretical part of the theory's language. To immerse yourself fully in the theoretical world-picture, hence to use the full theoretical language without qualm, branded one (once the irreducibility of theoretical terms was realized) with complete commitment to the veracity of that picture.

In the constructive empiricist alternative I have been developing, nothing is more natural, or more to be recommended than this total immersion. For the empirical import of the theory is now defined from within science, by means of a distinction between what is observable and what is not observable drawn by science itself. The epistemic commitment to the empirical import of the theory above (its empirical adequacy) can be stated using the language of science— and indeed, in no other way. It may be the case that I have no adequate way to describe this box, and the role it plays in my world, except as a VHF receiver. From this it does not follow that I believe that the concept of very high frequency electromagnetic waves corresponds to an individually identifiable element of reality. Concepts involve theories and are inconceivable without them, to paraphrase Sellars. But immersion in the theoretical world-picture does not preclude 'bracketing' its ontological implications.

After all, what is this world in which I live, breathe and have my being, and which my ancestors of two centuries ago could not enter? It is the intentional correlate of the conceptual framework through which I perceive and conceive the world. But our conceptual framework changes, hence the intentional correlate of our conceptual framework changes—but the real world is the same world.

What I have just said, denies what is called conceptual relativism. To be more precise, it denies it on this level, that is, on the level on which we interpret science and describe its role in our intellectual and practical life. Philosophy of science is not metaphysics—there may or may not be a deeper level of analysis on which that concept of the real world is subjected to scrutiny and found itself to be ... what? I leave to others the question whether we can consistently and coherently go further with such a line of thought. Philosophy of science can surely stay nearer the ground.

Let us discuss then the notion of *objectivity* as it appears in science. To someone immersed in that world-picture, the distinction between *electron* and *flying horse* is as clear as between *racehorse* and *flying horse*: the first corresponds to something in the actual world, and the other does not. While immersed in the theory, *and* addressing oneself solely to the problems in the domain of the theory, this objectivity of *electron* is not and cannot be qualified. *But this is so whether or not one is committed to the truth of the theory.* It is so not only for someone who believes, full stop, that the theory is true, but also for a Bayesian who grants a degree of belief equal to 1 to tautologies alone, and also to someone who is not a Bayesian but holds commitment to the truth of the theory in abeyance. For to say that someone is immersed in theory, 'living' in the theory's world, is not to describe his epistemic commitment. And if he describes his own epistemic commitment, he is stepping back for a moment, and saying something like: the theory entails that electrons exist, *and* not all theories do, *and* my epistemic attitude towards this theory is X.

We cannot revert to an earlier world-picture, because so many experimental findings cannot be accommodated in the science of an earlier time. This is not an argument for the truth of the present world-picture, but for its empirical adequacy. It is, you may wish to say, indirect or partial evidence for its truth—but only by being evidence for its empirical adequacy. It is also an argument for learning to find our way around in the world depicted by contemporary science, of speaking its language like a native. Someone who learns a second language comes to an all-important transition at a certain point: when he stops speaking by translating from his first language, and begins to speak 'directly'. It is only then that he begins to have access to the nuances and intangible differences that distinguish the two languages. The transition is a leap of sorts, into an unknown of sorts.

Not only objectivity, however, but also observability, is an intra-scientific distinction, if the science is taken wide enough. For that reason, it is possible even after total immersion in the world of science to distinguish possible epistemic attitudes to science, and to state them, and to limit one's epistemic commitment while remaining a functioning member of the scientific community—one who is reflective, and philosophically autonomous as well.

In my opinion, the phenomenology of science can be adequately discussed within the pragmatic analysis of language, to which I shall turn briefly at several points below.

§3. *The Conjunction Objection*

There have been arguments both for and against the idea that science as a whole aims at unity, that the development of a final single, coherent, consistent account incorporating all the special sciences is a regulative ideal governing the scientific enterprise. To some this seems a truism, to others it is mainly propaganda for the imperialism of physics (as they see it), and some have been concerned to point to the empirical presuppositions such an ideal may have.

Whatever the answer is at that grand level of debate, some sort of presumption of unity is pervasive in scientific practice. Scientists often use together theories that were developed originally for disparate domains of phenomena—chemistry and mechanics, mechanics and optics, physics and astronomy, chemistry and physiology. Sometimes such fields of joint enterprise receive special names: physical chemistry, molecular biology.

This conjunction of theories may seem the most obvious, uncontroversial of practices, but various realists (I think Putnam was the first) have pressed the objection that this practice is not intelligible on an anti-realist view.[5] Briefly, if one believes both T and T' to be true, then of course (on pain of inconsistency) one believes their conjunction to be true. But if T and T' are theories which are both empirically adequate, their conjunction need not be—it may even be inconsistent. Two rival theories, giving incompatible accounts of unobservable processes, could in principle each be empirically adequate.

Two things are clear about the objection as I have now presented it: that it rests on a logical point about empirical adequacy and truth, and that it must be made more precise. The latter is clear because in practice scientists must be very careful about conjoining theories, simply because even if (as the realists hold) acceptance is belief, it rarely happens that non-tentative, unqualified acceptance is warranted. There was a time when the Bohr–Sommerfeld theory of the atom looked victorious, and at the same time special relativity was generally accepted but—you see the point—it would not have made sense to conjoin them: the former would have to be subjected to a relativistic correction. Putting the two theories together would not have meant conjunction, but correction. Let me therefore first explore, briefly, the logical point behind the objection, and then discuss precise forms in which the conjunction objection may be raised.

A theory, on the simple view in which it is regarded as a body of statements, is true exactly if each of those statements is true. Indeed, each statement A can be regarded as a little theory, and there is a family $F(A)$ of models in which A is true. The family of models $F(T)$ in which T is true, consists of exactly those models that belong to $F(A)$ for each statement A which is part of T (or, which is implied by T). Logic is the study of *functions*, leading from statements (regarded as premisses) to other statements (regarded as conclusions) *which preserve truth*. Because of that intimate relation between the truth of a theory and the truth of its comprised statements, the logic of statements we all love and study gives rise to a logic of theories as well, in a perfectly straightforward manner.

Unlike truth, empirical adequacy is a global property of theories. There is no characteristic of statements such that, if all propositions of a theory individually have that characteristic, then the theory is empirically adequate. This point cannot be made precise unless we leave the simple view of theories used in the preceding paragraph, and think of a theory as a specification of a special family of models, each with a designated family of substructures that are meant to correspond to possible phenomena (empirical substructures). Of course, each statement that may be called a proposition of the theory is true in all these models, and each statement which cannot be called a proposition of the theory is false in at least one of these models. But since the empirical import of a theory cannot be syntactically isolated, we must define empirical adequacy directly, without an empirical detour: all the actual observable phenomena fit the empirical substructures in a certain one of these models.

It would not make sense therefore to ask for a discussion of the empirical adequacy of single statements, or for a logic of syntactic functions from premisses to conclusions which preserve empirical adequacy. About a single statement A, with family of models $F(A)$, the question of empirical adequacy can only be raised *with reference to a specific theory* T: does $F(A)$ include at least one of the models in the family specified by T that has this privileged status *vis-à-vis* the actual world? Unlike in the case of truth, the answer could be *yes* for one theory and *no* for another, with respect to the same statement A. Hence such a question, though construable, has no independent significance.

The process of putting theories together, in joint projects of explanation, prediction, and control, is a process the philosopher of science must be able to describe. In its first form, the conjunction objection draws a very simple picture of this process, by pointing to the logical rule

A

B

Therefore, A & B.

If a scientist believes theories T and T' to be true, then that explains why he uses them in conjunction, without a second thought, for just by being logical he will *a fortiori* believe that their conjunction is true—that is the putative explanation.

There can be no phenomena of the scientific life of which this simple account draws a faithful picture. The reason is that, as long as we are scientific in spirit, we cannot become dogmatic about even those theories which we whole-heartedly believe to be true. Hence a scientist must always, even if tacitly, reason *at least* as follows in such a case: if I believe T and T' to be true, then I also believe that (T and T') is true, and hence that it is empirically adequate. But in this new area of application T and T' are genuinely being used in conjunction; therefore, I will have a chance to see whether (T and T') really is empirically adequate, as I believe. That belief is not supported yet to the extent that my previous evidence supports the claims of empirical adequacy for T and T' separately, even though that support has been as good as I could wish it to be. Thus my beliefs are about to be put to a more stringent test in this joint application than they have ever been before.

What I have just described is the nearest practice can come to the simple account of putting theories together as mere conjunction. In my opinion, practice does not come that near; the preamble about believing T and T' to be true is missing, and what is about to be put to a more stringent test is the hypothesis that (T and T') is empirically adequate. This hypothesis is significant only, of course, if it is at least logically possible, that is, if T and T' have models in common—a theoretical question which should surely be taken seriously before any joint application is attempted.

But we may now raise the conjunction objection in another, more abstract form. Why should scientists want a single theory to cover disparate domains of phenomena, rather than a different, empirically adequate theory for each such domain? For the realist the motive is clear; for a theory cannot be true unless it can be *extended* consistently, without correction, to all of nature. But surely it is possible to have a lot of theories, each with its individual sorts of models, more or less overlapping in their domains of application—all empirically adequate, but impossible to combine into a single picture?

Pierre Duhem, himself a paradigm anti-realist, in his complaints about the broad but shallow mind of the English (as opposed to the deep but narrow mind of the French) actually charged that English physicists were content with such a piecemeal approach to the modelling of nature. If he is right, and if those physicists were genuine scientists, we should perhaps not be too anxious to explain that supposed regulative ideal of the unity of science. The explanation might at some point come to look rather like a representationalist theory of art, to which almost all twentieth-century art is an exception.

But we need not contemplate such *outré* possibilities, for it seems to me that the idea of a science consisting of a family of such disparate theories is really not feasible, except in the philosophically innocuous sense in which it actually does. Suppose, for example, that we try to have a mechanics and also a theory of electromagnetism, but no theory in which the two sorts of phenomena are both described. Where shall we find a home then for the phenomena involving moving charges? Electromagnetism would have to be electrostatics only. How could we have a successful physiology that does not take into account the effect of gravity which requires tensing of different muscles in different postures? One might contemplate teaching one theory of gravity to physiologists and another to astronomers. But at some point, someone will have to devise an account of the behaviour of the complex system consisting of a man in a space suit walking on the surface of the moon. Unless the two theories have models in common available for this situation, he must either conclude that the whole situation is impossible or else set about devising a theory that covers both the inanimate mechanism and the organism to be kept alive by it in the extraordinary condition of lunar gravity.

There remains then only the problem of living with lots of 'mini-theories' in practice, as we actually do. Physiologists need not make relativistic corrections in their mechanical calculations, and can treat almost all processes deterministically (and some stochastically which physics implies to be near deterministic). Philosophy of science could do with a more accurate picture of this situation—it is the actual situation of the working scientist and may well harbour problems obscured by our preoccupation with global theories.[6] But there seems to me no doubt that the aim of empirical adequacy already requires the successive unification of 'mini-theories' into larger ones, and that the process of unification is mainly one of correction and not of conjunction.

§4. *Pragmatic Virtues and Explanation*

§4.1 *The Other Virtues*

When a theory is advocated, it is praised for many features other than empirical adequacy and strength: it is said to be mathematically elegant, simple, of great scope, complete in certain respects: *also* of wonderful use in unifying our account of hitherto disparate phenomena, and most of all, explanatory. Judgements of simplicity and explanatory power are the intuitive and natural vehicle for expressing our epistemic appraisal.[7] What can an empiricist make of these other virtues which go so clearly beyond the ones he considers pre-eminent?

There are specifically human concerns, a function of our interests and pleasures, which make some theories more valuable or appealing to us than others. Values of this sort, however, provide reasons for using a theory, or contemplating it, whether or not we think it true, and cannot rationally guide our epistemic attitudes and decisions. For example, if it matters more to us to have one sort of question answered rather than another, that is no reason to think that a theory which answers more of the first sort of questions is more likely to be true (not even with the proviso 'everything else being equal'). It is merely a reason to prefer that theory in another respect.

Nevertheless, in the analysis of the appraisal of scientific theories, it would be a mistake to overlook the ways in which that appraisal is coloured by contextual factors. These factors are brought to the situation by the scientist from his own personal, social, and cultural situation. It is a mistake to think that the terms in which a scientific theory is appraised are purely hygienic, and have nothing to do with any other sort of appraisal, or with the persons and circumstances involved.

Theory acceptance has a pragmatic dimension. While the only belief involved in acceptance, as I see it, is the belief that the theory is empirically adequate, *more than belief is involved.* To accept a theory is to make a commitment, a commitment to the further confrontation of new phenomena within the framework of that theory, a commitment to a research programme, and a wager that all relevant phenomena can be accounted for without giving up that theory. That is why someone who has accepted a certain theory, will henceforth answer questions *ex cathedra*, or at least feel called upon to do so. Commitments are not true or false; they are vindicated or not vindicated in the course of human history.

Briefly, then, the answer is that the other virtues claimed for a theory are *pragmatic* virtues. In so far as they go beyond consistency, empirical adequacy, and empirical strength, they do not concern the relation between the theory and the world, but rather the use and usefulness of the theory; they provide reasons to prefer the theory independently of questions of truth.

Of course, this answer raises immediately the further question: why is this a *rational* procedure to follow in the appraisal of theories, in the deliberation that leads us to follow one approach rather than another in scientific research, or to commit ourselves epistemically, by accepting one theory rather than another?

I shall broach this question in the specific instance: why is it rational to pursue explanation? To answer this question fully we need an account of what explanation is—and I shall devote the next chapter to that. But beforehand, it is possible to sketch the answer which that account is meant to substantiate. It is this: the *epistemic* merits a theory may have or must have to figure in good explanations are not *sui generis*; they are just the merits it had in being empirically adequate, of significant empirical strength, and so forth. This does not mean that something is automatically a good explanation if it has those other merits; what more it needs is the pragmatic aspect of explanation. But in the pursuit of explanation we pursue *a fortiori* those more basic merits, which is what makes the pursuit of explanation of value to the scientific enterprise as such.

To praise a theory for its great explanatory power, is therefore to attribute to it *in part* the merits needed to serve the aim of science. It is not tantamount to attributing to it *special* features which make it more likely to be true, or empirically adequate. But it might be arguable that, for purely pragmatic (that is, person- and context-related) reasons, the pursuit of explanatory power is the best means to serve the central aims of science.

§4.2 *The Incursion of Pragmatics*

To spell out these contentions to the extent we can before we have an account of what explanation is, I must refer first of all to the terminology originally introduced by Charles Morris.[8] His basic concern was language, but we can transpose his concepts from words and statements to theories. In the study of language he saw three main levels: *syntax, semantics,* and *pragmatics*. The syntactic properties of an expression are determined only by its relations to other expressions, considered independently of meaning or interpretation. An example would be 'has six letters' which can be predicated of 'Cicero'. Semantic properties concern the relation of the expression to the world; an example is

1. 'Cicero' denotes Cicero.

Finally, pragmatics concerns the relation of the language to the users of that language; as in

2. Cicero preferred to be called 'Cicero' rather than 'Tully'.

In some sense, semantics is only an abstraction from pragmatics. It would not make sense to say 'I know that this man was named "Cicero" by his parents, and everyone always calls him that—but is his name really "Cicero"?' Yet we can study properties construed by abstracting from usage and its possible variations; this is merely one instance of scientific model building, in this case in the study of language.

But in certain cases, no abstraction is possible without losing the very thing we wish to study. How does the word 'I' differ from the word 'Cicero'? Exactly in the fact that the denotation of 'I' depends on who is using it—for every user uses it to refer to him or herself. Thus the semantic study of language can only go so far—then it must give way to a less thorough abstraction (that is, a less shallow level of analysis) and we find that we are doing pragmatics proper.

In the case of a statement, *truth* is the most important semantic property. A statement is true exactly if the actual world accords with this statement. But if some of the words, or grammatical devices, in that statement have a context-dependent semantic role, truth *simpliciter* does not make sense, and we must move again to pragmatics:

3. 'Cicero is dead' is true, if and only if Cicero is dead.
4. In any context or occasion of language use, 'I am happy' is true if and only if the person who says it on that occasion, is happy at the time of saying it.

Syntactic properties of and relations among statements include those studied in traditional logic, for 'is a logical truth', 'is not self-contradictory', 'is deducible from' are there all syntactically definable for large, useful fragments of our language.

Turning now to theories, we find there also a threefold division of properties and relations. First there are the purely internal or logical ones, such as axiomatizability, consistency, and various sorts of completeness. Attempts have been made to locate simplicity on this level, but these, as all other attempts so far to explain precisely what people could possibly mean when they call a theory simple or simpler, have failed.

Simplicity is quite an instructive case. It is obviously a criterion in theory choice, or at least a term in theory appraisal. For that reason, some writings on the subject of induction suggest that simple theories are more likely to be true. But it is surely absurd to think that the world is more likely to be simple than complicated (unless one has certain metaphysical or theological views not usually accepted as legitimate factors in scientific inference). The point is that the virtue, or patchwork of virtues, indicated by the term is a factor in theory appraisal, but does not indicate *special* features that make a theory more likely to be true (or empirically adequate).

Semantic properties and relations are those which concern the theory's relation to the world, or more specifically, the facts about which it is a theory. Here the two main properties are truth and empirical adequacy. Hence this is the area where both realism and constructive empiricism locate a central aim of science.

Are there also philosophically significant pragmatic theoretical properties? The working language of science is no doubt context-dependent, but surely that is a practical point only? Scientific theories can be stated in context-independent language, in what Quine calls 'eternal sentences'. So we do not need to stray into pragmatics, it would seem, to interpret science.

This may be true of those products of scientific activity which we call theories. It is not true of other parts of that activity, according to my view, and specifically I hold that

(a) the language of theory appraisal, and specifically the term 'explains' is radically context-dependent;
(b) the language of the use of theories to explain phenomena, is radically context-dependent.

These are two distinct points, for it is one thing to assert that Newton's theory explains the tides, and another to explain the tides by means of Newton's theory. For example, in doing the second you may never use the word 'explain'.

The pragmatics of language is also the place where we must locate such concepts as immersion in the language, or world-picture, of science. The basic factors in the linguistic situation, pragmatically conceived, are the speaker or user, the syntactic entity (sentence or set of sentences) uttered or displayed, the audience, and the factual circumstances. Any factor which relates to the speaker or audience is a pragmatic factor; and if it furthermore pertains specifically to that particular linguistic situation, a contextual factor. For example, the uttered word 'Cicero' may be discussed in isolation or in relation to the bearer of that name while still remaining on the level of abstraction properly called semantics. But the fact that it rather than 'Tully' was uttered, is a contextual factor. The fact that the speaker used it, on this occasion, to refer to his cat rather than to the senator, is also a contextual factor; that the speaker is a person in the habit of using the word that way, a pragmatic factor which may also play a role in this situation.

One such pragmatic or contextual factor may be a tacit agreement between speaker and audience (or unilateral commitment on the part of either) to be guided in his inferences by something more than bare logic. There may be a standing linguistic commitment, such as the educated layman's not to call anything table salt unless it is mainly sodium chloride, or the commitment, now falling into disuse, not to call anything cream unless it was produced by a cow. Such commitments may be more or less permanent or temporary. I know enough about astrology and about psychoanalysis to enter a conversation with an *aficionado* of either, in which that theory is what guides the use of terms and the allowed inferences. More commonly, the discussion of a movie, say, Renoir's *Day in the Country*, may go that way: 'Do you think he really seduced her? No, in that milieu a kiss was an earthshaking event.' A certain suspension of disbelief, a momentary commitment to the world depicted by the theory, play, painting, or novel, determines *in that linguistic situation* what it is correct to say, and the correct way to say it. Robert Stalnaker has given the name *pragmatic presuppositions* to propositions which play this role of guiding assumptions.

The total immersion in the scientific world-picture, which is proper to situations in which science is pursued or used, is a case in point. I shall again return to this topic at the end of Chapter 6 when discussing the use of modal language in science.

§4.3 *Pursuit of Explanation*

Very strong claims are sometimes made for the centrality of explanation among the aims of science. In some cases, indeed, the demand for explanation is held up as overriding and not subject to qualification, as unlimited. Such an extreme ideal of explanatory completeness we found in the arguments for scientific realism examined earlier on. But even more moderate philosophers, less easily accused of metaphysical leanings, make far-reaching claims. Thus, Ernest Nagel:

It is the desire for explanations which are at once systematic and controllable by factual evidence that generates science; and it is the organization and classification of knowledge on the basis of explanatory principles that is the distinctive goal of the sciences.[9]

None of this entails realism, and the first part is, I think, undoubtedly true. The second part might still conflict with the view that empirical adequacy is the pre-eminent virtue, depending on how 'explanatory' is understood. But if we look to how Nagel understands explanation, we find that he holds to an account that is rather like Hempel's (to be examined in the next chapter). Let us call whatever Nagel understands to be explanation, *N-explanation*. Then Nagel here reported his conviction that the distinctive goal of the sciences is *N*-explanation. This may well be true even if the search for explanation is totally explicable as being of value to science because it serves the aim of giving us empirically adequate and strong theories. That this is not a far-fetched construal, seems to me clear from the following passage on the same page, where he gives as main example for his contention that a few principles formulated by Newton

suffice to show that propositions concerning the moon's motion, the behavior of the tides, the paths of projectiles, and the rise of liquids in thin tubes *are intimately related*, and that all these propositions *can be rigorously deduced* from those principles conjoined with various special assumptions of fact.

This certainly does not contradict the idea that the name of the game is saving the phenomena, even while there is a strong flavour of that distinctive satisfaction the human mind finds in encompassing an elegant, tightly and coherently constructed theory in order to win in that game.

There is a totally false issue that tends to be brought up in this connection—and was, for instance, by Paul Feyerabend.[10] Let us suppose for a moment that what more there is to explanation is merely a function of human interests. Then scientists need not worry unduly about achieving explanation over and above empirical adequacy. They can stop when they believe that they have *that*. However, in the history of science it is clear that scientists would have been ill advised to be so sanguine. The search for a dynamics compatible with Copernicus's new astronomical scheme, the search for the details of the atomic structure that would explain discrete spectra, the pursuit of the kinetic theory even when phenomenological thermodynamics seemed entirely adequate—there are many examples in which the search for explanation paid off handsomely. So only Realism is a philosophy that stimulates scientific inquiry; anti-realism hampers it.

Paid off handsomely, how? Paid off in new theories we have more reason to believe empirically adequate. But in that case even the anti-realist, when asked questions about *methodology* will *ex cathedra* counsel the search for explanation! We might even suggest a loyalty oath for scientists, if realism is so efficacious. In any case, the criticism is based on a very naïve view of scientific certainty; there always have been reasons to doubt the empirical adequacy of extant theories and these were reasons operative in the cited examples of 'searches for explanation'.

I call this a false issue, for the interpretation of science, and the correct view of its methodology, are two separate topics. But I have sketched *en passant* my answer to the question about methodology: the search for explanation is valued in science because it consists *for the most part* in the search for theories which are simpler, more unified, and more likely to be empirically adequate. This is not because explanatory power is a separate quality *sui generis* which, mysteriously, makes those other qualities more likely, but because having a good explanation *consists* for the most part in having a theory with those other qualities.

To see whether explanation really is pre-eminent among the theoretical virtues sought in science, we should gauge how it is regarded in competition with other virtues.

First, there are rock-bottom criteria of minimal acceptability: consistency, internally and with the facts. Cases are known of mathematically inconsistent theories (Dirac introduced a function at one point, which was very useful, but later shown to be an impossible one) but that is a defect that must be repaired. You cannot advocate a theory as correct and inconsistent. Inconsistency with the observed facts is similarly minimal; if the theory conflicts with any previously acceptable data, we must either change the theory or deny that those data are correct.

Explanation is not a rock-bottom minimal virtue of this sort. If explanation of the facts were required in the way consistency with the facts is, then every theory would have to explain every fact in its domain. Newton would have had to add an explanation of gravity to his celestial mechanics before presenting it at all. But instead he says:

Hitherto we have explained the phaenomena of the heavens and of our sea, by the power of Gravity, but have not yet assign'd a cause of this power ... hitherto I have not been able to discover the cause of those properties of gravity from phaenomena, and I frame no hypotheses. For whatever is not deduc'd from the phaenomena, is to be called an hypothesis; and hypotheses, whether metaphysical or physical, whether of occult qualities or mechanical, have no place in experimental philosophy ... And to us it is enough, that gravity does really exist, and act according to the laws which we have explained, and abundantly serves to account for all the motions of the celestial bodies, and of our sea.[11]

His reasons are perhaps stronger than many would accept, but it remains that he can decline to explain whereas he couldn't very well decline to be consistent. Another example is that Newton did decline, clearly, to satisfy a criterion of Kepler for the adequacy of any theory of the heavens: that it should explain why there are exactly six planets.[12]

The question is then whether explanatory power would be preferred over other virtues when there is a conflict. This can surely not be so when the other virtue is the one the non-realist regards as the highest—empirical adequacy. For to forgo empirical adequacy is to allow that there may arise inconsistencies with observed facts. That possibility we cannot allow while advocating a theory as correct. Indeed, empirical adequacy is a precondition: we don't say that we have an explanation unless we have an *acceptable* theory which explains.

Thirdly, we may ask whether explanation is a pre-eminent virtue in the sense of being required when it can be had. This would mean that if several theories were empirically equivalent, the one which explains most would have to be accepted. Against this idea count all the examples of scientists refusing to enlarge their theories in ways that do not yield different (or further) empirical consequences. One example is already given by the passage I quoted from Newton, but another instructive example is yet another feature of the discussion of hidden variables for quantum mechanics.

According to quantum theory there are correlations in the behaviour of particles which have interacted in the past, but are now physically separated. No causal mechanism is given to explain these correlations, which were dramatized in a famous paper by Einstein, Podolski, and Rosen. Various experiments have borne out the existence of these correlations, which may be found for instance if an atom 'cascading down' from an excited state, emits two photons. When polarization filters are set up for these photons to pass through, it is as if each photon 'knows' whether the other photon passed the other filter.

Certain hidden variable theories have been proposed which would explain such correlations (so-called 'hidden variable theories of the second kind').[13] These do not predict exactly the same correlations—this is what makes these theories interesting to physics. So far, experiments appear to support quantum theories against those rivals. But the one response which is conspicuous by its absence is that an explanation of the correlations *must be found* which fits in exactly with quantum theory and does not affect its empirical content at all. Such metaphysical extensions of the theory (if indeed possible) would be philosophical playthings only. There are only two camps to the debate as far as physics is concerned: either this non-locality makes quantum theory pre-eminently suited to the representation of the world (and we need to re-school our imaginations), or else quantum theory must be replaced by an empirically significant *rival*.

In none of the three senses examined is explanation an overriding virtue. Philosophy fathered the sciences, and philosophy aims pre-eminently, and perhaps only, to remove wonder, as Aristotle said; but these children have left the parental home.

The Pragmatics of Explanation[1]

> If cause were non-existent everything would have been pro-
> duced by everything and at random. Horses, for instance,
> might be born, perchance, of flies, and elephants of ants;
> and there would have been severe rains and snow in Egyp-
> tian Thebes, while the southern districts would have had
> no rain, unless there had been a cause which makes the
> southern parts stormy, the eastern dry.
> Sextus Empiricus, *Outlines of Pyrrhonism*
> III, V, 1

A THEORY is said to have explanatory power if it allows us to explain;
and this is a virtue. It is a pragmatic virtue, albeit a complex one
that includes other virtues as its own preconditions. After some pre-
liminaries in Section 1, I shall give a frankly selective history of philosophical attempts to explain explanation. Then I shall offer a model
of this aspect of scientific activity in terms of why-questions, their
presuppositions, and their context-dependence. This will account for
the puzzling features (especially asymmetries and rejections) that
have been found in the phenomenon of explanation, while remaining
compatible with empiricism.

§1. *The Language of Explanation*

One view of scientific explanation is encapsulated in this argument:
science aims to find explanations, but nothing is an explanation unless it is true (explanation requires true premises); so science aims
to find true theories about what the world is like. Hence scientific
realism is correct. Attention to other uses of the term 'explanation'
will show that this argument trades on an ambiguity.

§1.1 *Truth and Grammar*

It is necessary first of all to distinguish between the locutions 'we
have an explanation' and 'this theory explains'. The former can be
paraphrased 'we have a theory that explains'—but then 'have' needs

to be understood in a special way. It does not mean, in this case, 'have on the books', or 'have formulated', but carries the conversational implicature that the theory tacitly referred to is acceptable. That is, you are not warranted in saying 'I have an explanation' unless you are warranted in the assertion 'I have a theory *which is acceptable* and which explains'. The important point is that the mere statement 'theory T explains fact E' does not carry any such implication: not that the theory is true, not that it is empirically adequate, and not that it is acceptable.

There are many examples, taken from actual usage, which show that truth is not presupposed by the assertion that a theory explains something. Lavoisier said of the phlogiston hypothesis that it is too vague and consequently 's'adapte à toutes les explications dans lesquelles on veut le faire entrer'.[2] Darwin explicitly allows explanations by false theories when he says 'It can hardly be supposed that a false theory would explain, in so satisfactory a manner as does the theory of natural selection, the several large classes of facts above specified.'[3] Gilbert Harman, we recall, has argued similarly: that a theory explains certain phenomena is part of the evidence that leads us to accept it. But that means that the explanation-relation is visible before we believe that the theory is true. Finally, we criticize theories selectively: a discussion of celestial mechanics around the turn of the century could surely contain the assertion that Newton's theory does explain many planetary phenomena. Yet it was also agreed that the advance in the perihelion of Mercury seems to be inconsistent with the theory, suggesting therefore that the theory is not empirically adequate—and hence, is false—without this agreement undermining the previous assertion. Examples can be multiplied: Newton's theory explained the tides, Huygens's theory explained the diffraction of light, Rutherford's theory of the atom explained the scattering of alpha particles, Bohr's theory explained the hydrogen spectrum, Lorentz's theory explained clock retardation. We are quite willing to say all this, although we will add that, for each of these theories, phenomena were discovered which they could not only not explain, but could not even accommodate in the minimal fashion required for empirical adequacy.

Hence, to say that a theory explains some fact or other, is to assert a relationship between this theory and that fact, which is independent of the question whether the real world, as a whole, fits that theory.

Let us relieve the tedium of terminological discussion for a moment and return to the argument displayed at the beginning. In view of the distinctions shown, we can try to revise it as follows: science tries to place us in a position in which we have explanations, and are warranted in saying that we do have. But to have such warrant, we must first be able to assert with equal warrant that the theories we use to provide premisses in our explanations are true. Hence science tries to place us in a position where we have theories which we are entitled to believe to be true.

The conclusion may be harmless of course if 'entitled' means here only that one can't be convicted of irrationality on the basis of such a belief. That is compatible with the idea that we have warrant to believe a theory only because, and in so far as, we have warrant to believe that it is empirically adequate. In that case it is left open that one is at least as rational in believing merely that the theory is empirically adequate.

But even if the conclusion were construed in this harmless way, the second premiss will have to be disputed, for it entails that someone who merely accepts the theory as empirically adequate, is not in a position to explain. In this second premiss, the conviction is perhaps expressed that having an explanation is not to be equated with having an acceptable theory that explains, but with having a true theory that explains.

That conviction runs afoul of the examples I gave. I say that Newton could explain the tides, that he had an explanation of the tides, that he did explain the tides. In the same breath I can add that this theory is, after all, not correct. Hence I would be inconsistent if by the former I meant that Newton had a true theory which explained the tides—for if it was true then, it is true now. If what I meant was that it was true *then* to say that Newton had an acceptable theory which explains the tides, that would be correct.

A realist can of course give his own version: to have an explanation means to have 'on the books' a theory which explains and which one is entitled to believe to be true. If he does so, he will agree that to have an explanation does not require a true theory, while maintaining his contention that science aims to place us in a position to give *true* explanations. That would bring us back, I suppose, to our initial disagreement, the detour through explanation having brought no benefits. If you can only be entitled to assert that the theory is true because, and in so far as, you are entitled to assert that it is empirically adequate, then the distinction drawn makes no practical difference. There would of course be a difference between *believe* (to-be-true) and *accept* (believe-to-be-empirically-adequate) but no real difference between be-entitled-to-believe and be-entitled-to-accept. A realist might well dispute this by saying that if the theory explains facts then that gives you an *extra* good reason (over and above any evidence that it is empirically adequate) to believe that the theory is true. But I shall argue that this is quite impossible, since explanation is not a special additional feature that can give you good reasons for belief in addition to evidence that the theory fits the observable phenomena. For 'what more there is to' explanation is something quite pragmatic, related to the concerns of the user of the theory and not something new about the correspondence between theory and fact.

So I conclude that (a) the assertion that theory *T* explains, or provides an explanation for, fact *E* does not presuppose or imply that *T* is true or even empirically adequate, and (b) the assertion that we have an explanation is most simply construed as meaning that we have 'on the books' an acceptable theory which explains. I shall henceforth adopt this construal.

To round off the discussion of the terminology, let us clarify what sorts of terms can be the grammatical subjects, or grammatical objects, of the verb 'to explain'. Usage is not regimented: when we say 'There is the explanation!', we may be pointing to a fact, or to a theory, or to a thing. In addition, it is often possible to point to more than one thing which can be called 'the explanation'. And, finally, whereas one person may say that Newton's theory of gravitation explained the tides, another may say that Newton used that theory to explain the tides. (I suppose no one would say that the hammer drove the nail through the wood; only that the carpenter did so, using the hammer. But today people do sometimes say that the computer calculated the value of a function, or solved the equations, which is perhaps similar to saying that the theory explained the tides.)

This bewildering variety of modes of speech is common to scientists as well as philosophers and laymen. In Huygens and Young the typical phrasing seemed to be that phenomena may be explained *by means of* principles, laws, and hypotheses, or *according* to a view.[4] On the other hand, Fresnel writes to Arago in 1815 'tous ces phénomènes... sont réunis et expliqués par la même théorie des vibrations', and Lavoisier says that the oxygen hypothesis he proposes *explains* the phenomena of combustion.[5] Darwin also speaks in the latter idiom: 'In scientific investigations it is permitted to invent any hypothesis, and if it explains various large and independent classes of facts it rises to the rank of a well-grounded theory'; though elsewhere he says that the facts of geographical distribution are *explicable on* the theory of migration.[6]

In other cases yet, the theory assumed is left tacit, and we just say that one fact explains another. For example: the fact that water is a chemical compound of oxygen and hydrogen explains why oxygen and hydrogen appear when an electric current is passed through (impure) water.

To put some order into this terminology, and in keeping with previous conclusions, we can regiment the language as follows. The word 'explain' can have its basic role in expressions of form 'fact E explains fact F relative to theory T'. The other expressions can then be parsed as: 'T explains F' is equivalent to: 'there are facts which explain F relative to T'; 'T was used to explain F' equivalent to 'it was shown that there are facts which explain F relative to T'; and so forth. Instead of 'relative to T' we can sometimes also say 'in T'; for example, 'the gravitational pull of the moon explains the ebb and flow of the tides in Newton's theory'.

After this, my concern will no longer be with the derivative type of assertion that we *have* an explanation. After this point, the topic of concern will be that basic relation of explanation, which may be said to hold between facts relative to a theory, quite independently of whether the theory is true or false, believed, accepted, or totally rejected.

§1.2 *Some Examples*

Philosophical discussion is typically anchored to its subject through just a few traditional examples. The moment you see 'Pegasus', 'the king of France', or 'the good Samaritan', in a philosophical paper, you know exactly to what problem area it belongs. In the philosophical discussion of explanation, we also return constantly to a few main examples: paresis, the red shift, the flagpole. To combat the increasing sense of unreality this brings, it may be as well to rehearse, briefly, some workaday examples of scientific explanation.

(1) Two kilograms of copper at 60° C are placed in three kilograms of water at 20° C. After a while, water and copper reach the same

temperature, namely 22.5° C, and then cool down together to the temperature of the surrounding atmosphere.

There are a number of facts here for which we may request an explanation. Let us just ask why the equilibrium temperature reached is 22.5° C.

Well, the specific heats of water and copper are 1 and 0.1, respectively. Hence if the final temperature is T, the copper loses $0.1 \times 2 \times (60-T)$ units of heat and the water gains $1 \times 3 \times (T-20)$. At this point we appeal to the principle of the Conservation of Energy, and conclude that the total amount of heat neither increased nor diminished. Hence,

$$0.1 \times 2 \times (60-T) = 1 \times 3 \times (T-20)$$

from which $T=22.5$ can easily be deduced.

(2) A short circuit in a power station results in a momentary current of 10^6 amps. A conductor, horizontally placed, 2 metres in length and 0.5 kg in mass, is warped at that time.

Let us ask why the conductor was warped. Well, the earth's magnetic field at this point is not negligible; its vertical component is approximately $5/10^5$ tesla. The theory of electro-magnetism allows us to calculate the force exerted on the conductor at the time in question:

$$(5/10^5) \times 2 \times 10^6 = 100 \text{ newtons}$$

which is directed at right angles to the conductor in the horizontal plane. The second law of Newton's mechanics entails in turn that, at that moment, the conductor has an acceleration of

$$100 \div 0.5 = 200 \text{ m/sec}^2$$

which is approximately twenty times the downward acceleration attributable to gravity (9.8 m/sec^2)—which allows us to compare in concrete terms the effect of the short circuit on the fixed conductor, and the normal effect of its weight.

(3) In a purely numerological way, Balmer, Lyman, and Paschen constructed formulae fitting frequency series to be found in the hydrogen spectrum, of the general form:

$$f_m^n = R\left(\frac{1}{m^2} - \frac{1}{n^2}\right)$$

where Balmer's law had $m=2$, Lyman's had $m=1$, and Paschen's $m=3$; both m and n range over natural numbers.

Bohr's theory of the atom explains this general form. In this theory, the electron in a hydrogen atom moves in a stable orbit, characterized by an angular momentum which is an integral multiple of $h/2\pi$. The associated energy levels take the form

$$E_n = -E_o/n^2$$

where E_o is called the ground state energy.

When the atom is excited (as when the sample is heated), the electron jumps into a higher energy state. It then spontaneously drops down again, emitting a photon with energy equal to the energy lost by that electron in its drop. So if the drop is from level E_n to level E_m, the photon's energy is

$$E = E_n - E_m = (-E_o/n^2) - (-E_o/m^2)$$
$$= E_o/m^2 - E_o/n^2$$

The frequency is related to the energy by the equation

$$E = hf$$

so the frequencies exhibited by the emitted photons are

$$f_m^n = \frac{E}{h} = \frac{E_o}{h}\left(\frac{1}{m^2} - \frac{1}{n^2}\right)$$

which is exactly of the general form found above, with E_o/h being the constant R.

The reader may increase this stock of examples by consulting elementary texts and the *Science Digest*. It should be clear at any rate that scientific theories are used in explanation, and that how well a theory is to be regarded, depends at least in part on how much it can be used to explain.

§2. *A Biased History*

Current discussion of explanation draws on three decades of debate, which began with Hempel and Oppenheim's 'Studies in the Logic of Explanation' (1948).[7] The literature is now voluminous, so that a retrospective must of necessity be biased. I shall bias my account in such a way that it illustrates my diagnoses of the difficulties and points suggestively to the solution I shall offer below.

§2.1 *Hempel: Grounds for Belief*

Hempel has probably written more papers about scientific explanation than anyone; but because they are well known I shall focus on the short summary which he gave of his views in 1966.[8] There he lists two criteria for what is an explanation:

explanatory relevance: 'the explanatory information adduced affords good grounds for believing that the phenomenon did, or does, indeed occur.'

testability: 'the statements constituting a scientific explanation must be capable of empirical test.'

In each explanation, the information adduced has two components, one ('the laws') information supplied by a theory, and the other ('the initial or boundary conditions') being auxiliary factual information. The relationship of providing good grounds is explicated separately for statistical and non-statistical theories. In the latter, the information *implies* the fact that is explained; in the former, it *bestows high probability* on that fact.

As Hempel himself points out, the first criterion does not provide either sufficient or necessary conditions for explanation. This was established through a series of examples given by various writers (but especially Michael Scriven and Sylvain Bromberger) and which have passed into the philosophical folklore.

First, giving good grounds for belief does not always amount to explanation. This is most strikingly apparent in examples of the asymmetry of explanation. In such cases, two propositions are strictly equivalent (relative to the accepted background theory), and the one can be adduced to explain why the other is the case, but not conversely. Aristotle already gave examples of this sort (*Posterior Analytics*, Book I, Chapter 13). Hempel mentions the phenomenon of the *red shift*: relative to accepted physics, the galaxies are receding from us if and only if the light received from them exhibits a shift toward the red end of the spectrum. While the receding of the galaxies can be cited as the reason for the red shift, it hardly makes sense to say that the red shift is the reason for their motion. A more simple-minded example is provided by the *barometer*, if we accept the simplified hypothesis that it falls exactly when a storm is coming, yet does not explain (but rather, is explained by) the fact that a storm is coming. In both examples, good grounds of belief are provided by either proposition for the other. The flagpole is perhaps the most famous asymmetry. Suppose that a flagpole, 100 feet high, casts a shadow 75 feet long. We can explain the length of the shadow by noting the angle of elevation of the sun, and appealing to the accepted theory that light travels in straight lines. For given that angle, and the height of the pole, trigonometry enables us to deduce the length of the base of the right-angled triangle formed by pole, light ray, and shadow. However, we can similarly deduce the length of the pole from the length of the shadow plus the angle of elevation. Yet if someone asks us why the pole is 100 feet high, we cannot explain that fact by saying 'because it has a shadow 75 feet long'. The most we could explain that way is how we *came to know*, or how he might himself verify the claim, that the pole is indeed so high.

Second, not every explanation is a case in which good grounds for belief are given. The famous example for this is *paresis*: no one contracts this dreadful illness unless he had latent, untreated syphilis. If someone asked the doctor to explain to him why he came down with this disease, the doctor would surely say: 'because you had latent syphilis which was left untreated'. But only a low percentage of such cases are followed by paresis. Hence if one knew of someone that he might have syphilis, it would be reasonable to warn him that, if left untreated, he might contract paresis—but not reasonable to expect him to get it. Certainly we do not have here the high probability demanded by Hempel.

It might be replied that the doctor has only a partial explanation, that there are further factors which medical science will eventually discover. This reply is based on faith that the world is, for macroscopic phenomena at least, deterministic or nearly so. But the same point can be made with examples in which we do not believe that there is further information to be had, even in principle. The half-life of uranium U^{238} is $(4.5) \cdot 10^9$ years. Hence the probability that a given small enough sample of uranium will emit radiation in a specified small interval of time is low. Suppose, however, that it does. We still say that atomic physics explains this, the explanation being that this material was uranium, which has a certain atomic structure, and hence is subject to spontaneous decay. Indeed, atomic physics has many more examples of events of very low probability, which are explained in terms of the structure of the atoms involved. Although there are physicists and philosophers who argue that the theory must therefore be incomplete (one of them being Einstein, who said 'God does not play with dice') the prevalent view is that it is a contingent matter whether the world is ultimately deterministic or not.

In addition to the above, Wesley Salmon raised the vexing problem of *relevance* which is mentioned in the title of the first criterion, but does not enter into its explication. Two examples which meet the requirements of providing good grounds are:

John Jones was almost certain to recover from his cold because he took vitamin C, and almost all colds clear up within a week of taking vitamin C.

John Jones avoided becoming pregnant during the past year, for he has taken his wife's birth control pills regularly, and every man who takes birth control pills avoids pregnancy.[9]

Salmon assumed here that almost all colds spontaneously clear up within a week. There is then something seriously wrong with these 'explanations', since the information adduced is wholly or partly irrelevant. So the criterion would have to be amended at least to read: 'provides good and *relevant* grounds'. This raises the problem of explicating relevance, also not an easy matter.

The second criterion, of testability, is met by all scientific theories, and by all the auxiliary information adduced in the above examples, so it cannot help to ameliorate these difficulties.

§2.2 *Salmon: Statistically Relevant Factors*

A number of writers adduced independent evidence for the conclusion that Hempel's criterion is too strong. Of these I shall cite three. The first is Morton Beckner, in his discussion of evolution. This is not a deterministic theory, and often explains a phenomenon only by showing how it could have happened—and indeed, might well have happened in the presence of certain describable, believable conditions consistent with the theory.

Selectionists have devoted a great deal of effort to the construction of models that are aimed at demonstrating that some observed or suspected phenomena are possible, that is, that they are compatible with the established or confirmed biological hypotheses ... These models all state strongly that if conditions were (or are) so and so, then, the laws of genetics being what they are, the phenomena in question must occur.[10]

Thus evolution theory explains, for example, the giraffe's long neck, although there was no independent knowledge of food shortages of the requisite sort. Evolutionists give such explanations by constructing models of processes which utilize only genetic and natural selection mechanisms, in which the outcome agrees with the actual phenomena.

In a similar vein, Putnam argued that Newton's explanations were *not* deductions of the facts that had to be explained, but rather demonstrations of compatibility. What was demonstrated was that celestial motions could be as they were, given the theory and certain possible mass distributions in the universe.[11]

The distinction does not look too telling as long as we have to do with a deterministic theory. For in that case, the phenomena *E* are *compatible* with theory *T* if and only if there are possible preceding conditions *C* such that *C* plus *T imply E*. In any case, deduction and merely logical consistency cannot be what is at issue, since to show that *T* is logically compatible with *E* it would suffice to show that *T* is irrelevant to (has nothing to say about) *E*—surely not sufficient for explanation.

What Beckner and Putnam are pointing to are demonstrations that tend to establish (or at least remove objections to) claims of empirical adequacy. It is shown that the development of the giraffe's neck, or the fly-whisk tail fits a model of evolutionary theory; that the observed celestial motions fit a model of Newton's celestial mechanics. But a claim of empirical adequacy does not amount to a claim of explanation—there must be more to it.

Wesley Salmon introduced the theory that an explanation is not an argument, but an assembly of statistically relevant factors. A fact *A* is statistically relevant to a phenomenon *E* exactly if the probability of *E given A* is different from the probability of *E simpliciter*:

$$P(E/A) \neq P(E)$$

Hempel's criterion required $P(E/A)$ to be high (at least greater than $\frac{1}{2}$). Salmon does not require this, and he does not even require that the information *A* increases the probability of *E*. That Hempel's requirement was too strong, is shown by the paresis example (which fits Salmon's account very well), and that $P(E/A)$ should not be required to be higher than $P(E)$ Salmon argues independently.

He gives the example of an equal mixture of Uranium-238 atoms and Polonium-214 atoms, which makes the Geiger counter click in interval $(t, t+m)$. This means that one of the atoms disintegrated. Why did it? The correct answer will be: because it was a Uranium-238 atom; if that is so—although the probability of its disintegration is much higher relative to the previous knowledge that the atom belonged to the described mixture.[12] The problem with this argument is that, on Salmon's criterion, we can explain not only why there was a disintegration, but also why the disintegration occurred, let us say, exactly half-way between t and $t+m$. For the information is statistically relevant to that occurrence. Yet would we not say that this is the sort of fact that atomic physics leaves unexplained?

The idea behind this objection is that the information is statistically relevant to the occurrence at $t + (m/2)$, but does not favour that as against various other times in the interval. Hence, if $E = $ (a disintegration occurred) and $E_x = $ (a disintegration occurred at time x), then Salmon bids us compare $P(E_x)$ with $P(E_x/A)$, whereas we naturally compare *also* $P(E_x/A)$ with $P(E_y/A)$ for other times y. This suggests that mere statistical relevance is not enough.

Nancy Cartwright has provided several examples to show that Salmon's criterion of statistical relevance also does not provide necessary or sufficient conditions for explanation.[13] As to sufficiency, suppose I spray poison ivy with defoliant which is 90 per cent effective. Then the question 'Why is *this* poison ivy now dead?' may correctly be answered 'Because it was sprayed with the defoliant.' About 10 per cent of the plants are still alive, however, and for those it is true that the probability that they are still alive was not the same as the probability that they are still alive *given* that they were sprayed. Yet the question 'Why is *that* plant now alive?' cannot be correctly answered 'Because it was sprayed with defoliant.'

Nor is the condition necessary. Suppose, as a medical fiction, that paresis can result from either syphilis or epilepsy, and from nothing else, and that the probability of paresis given either syphilis or epilepsy equals 0.1. Suppose in addition that Jones is known to belong to a family of which every member has either syphilis or epilepsy (but, fortunately, not both), and that he has paresis. Why did *he* develop this illness? Surely the best answer *either* is 'Because he had syphilis' *or* is 'Because he had epilepsy', depending on which of these is true. Yet, with all the other information we have, the probability that Jones would get paresis is already established as 0.1, and this probability is not changed if we are told in addition, say, that he has a history of syphilis. The example is rather similar to that of the Uranium and Polonium atoms, except that the probabilities are equal—and we still want to say that in *this* case, *the* explanation of the paresis is the fact of syphilis.

Let me add a more general criticism. It would seem that if either Hempel's or Salmon's approach was correct, then there would not really be more to explanatory power than empirical adequacy and empirical strength. That is, on these views, explaining an observed event is indistinguishable from showing that this event's occurrence does not constitute an objection to the claim of empirical adequacy for one's theory, and in addition, providing significant information entailed by the theory and relevant to that event's occurrence. And it seems that Salmon, at that point, was of the opinion that there really cannot be more to explanation:

When an explanation ... has been provided, we know exactly how to regard any A with respect to the property B ... We know all the regularities (universal or statistical) that are relevant to our original question. What more could one ask of an explanation?[14]

But in response to the objections and difficulties raised, Salmon, and others, developed new theories of explanation according to which there is more to explanatory power. I shall examine Salmon's later theory below.

§2.3 *Global Properties of Theories*

To have an explanation of a fact is to have (accepted) a theory which is acceptable, and which explains that fact. The latter relation must indubitably depend on what that fact is, since a theory may explain one fact and not another. Yet the following may also be held: it is a necessary condition that the theory, considered as a whole, has certain features beyond acceptability. The relation between the theory and *this* fact may be called a *local* feature of the theory, and characters that pertain to the theory taken as a whole, *global* features.

This suggestive geometric metaphor was introduced by Michael Friedman, and he attempted an account of explanation along these lines. Friedman wrote:

On the view of explanation that I am proposing, the kind of understanding provided by science is global rather than local. Scientific explanations do not confer intelligibility on individual phenomena by showing them to be some-how natural, necessary, familiar, or inevitable. However, our overall under-standing of the world is increased...[15]

This could be read as totally discounting a specific relation of explanation altogether, as saying that theories can have certain over-all virtues, at which we aim, and because of which we may ascribe to them explanatory power (with respect to their primary domain of application, perhaps). But Friedman does not go quite as far. He gives an explication of the relation *theory T explains phenomenon P*. He supposes (p. 15) that phenomena, i.e. general uniformities, are represented by lawlike sentences (whatever those may be); that we have as background a set K of accepted lawlike sentences, and that the candidate S (law, theory, or hypothesis) for explaining P is itself representable by a lawlike sentence. His definition has the form:

S explains P exactly if P is a consequence of S, relative to K, and S 'reduces' or 'unifies' the set of its own consequences relative to K.

Here A is called a consequence of B relative to K exactly if A is a consequence of B and K together. He then modifies the above formula, and explicates it in a technically precise way. But as he explicates it, the notion of reduction cannot do the work he needs it to do, and it does not seem that anything like his precise definition could do.[16] More interesting than the details, however, is the form of the intuition behind Friedman's proposal. According to him, we evaluate something as an explanation relative to an assumed background theory K. I imagine that this theory might actually include some auxiliary information of a non-lawlike character, such as the age of the universe, or the boundary conditions in the situation under study. But of course K could not very well include all our information, since we generally know that P when we are asking for an explanation of P. Secondly, relative to K, the explanation implies that P is true. In view of Salmon's criticisms, I assume that Friedman would wish to weaken this Hempel-like condition. Finally, and here is the crux, it is the character of K plus the adduced information together, regarded as a complex theory, that determines whether we have an explanation. And the relevant features in this determination are global features, having to do with all the phenomena covered, not with P as such. So, whether or not K plus the adduced information provides new information about facts other than those described in P, appears to be crucial to whether we have an explanation of P.

James Greeno has made a similar proposal, with special reference to statistical theories. His abstract and closing statement says:

The main argument of this paper is that an evaluation of the overall explanatory power of a theory is less problematic and more relevant as an assessment of the state of knowledge than an evaluation of statistical explanations of single occurrences ...[17]

Greeno takes as his model of a theory one which specifies a single probability space Q as the correct one, plus two partitions (or random variables) of which one is designated *explanandum* and the other *explanans*. An example: sociology cannot explain why Albert, who lives in San Francisco and whose father has a high income, steals a car. Nor is it meant to. But it does explain delinquency in terms of such other factors as residence and parental income. The degree of explanatory power is measured by an ingeniously devised quantity which measures the information I the theory provides of the *explanandum* variable M on the basis of *explanans S*. This measure takes its maximum value if all conditional probabilities $P(M_i/S_j)$ are zero or one (D–N case), and its minimum value zero if S and M are statistically independent.

But it is not difficult to see that Greeno's way of making these ideas precise still runs into some of the same old difficulties. For suppose S and M describe the behaviour of barometers and storms. Suppose that the probability that the barometer will fall (M_1) equals the probability that there will be a storm (S_1), namely 0.2, and that the probability that there is a storm *given* that the barometer falls equals the probability that the barometer falls *given* that there will be a storm, namely 1. In that case the quantity I takes its maximum value—and indeed, does so even if we interchange M and S. But surely we do not have an explanation in either case.

§2.4 *The Difficulties: Asymmetries and Rejections*

There are two main difficulties, illustrated by the old paresis and barometer examples, which none of the examined positions can handle. The first is that there are cases, clearly in a theory's domain, where the request for explanation is nevertheless rejected. We can explain why John, rather than his brothers, contracted paresis, for he had syphilis; but not why he, among all those syphilitics, got paresis. Medical science is incomplete, and hopes to find the answer some day. But the example of the uranium atom disintegrating just then rather than later, is formally similar and we believe the theory to be complete. We also reject such questions as the Aristotelians asked the Galileans: why does a body free of impressed forces retain its velocity? The importance of this sort of case, and its pervasive character, has been repeatedly discussed by Adolf Grünbaum. It was also noted, in a different context, by Thomas Kuhn.[18] Examples he gives of explanation requests which were considered legitimate in some periods and rejected in others cover a wide range of topics. They include the qualities of compounds in chemical theory (explained before Lavoisier's reform, and not considered something to be explained in the nineteenth century, but now again the subject of chemical explanation). Clerk Maxwell accepted as legitimate the request to explain electromagnetic phenomena within mechanics. As his theory became more successful and more widely accepted, scientists ceased to see the lack of this as a shortcoming. The same had happened with Newton's theory of gravitation which did not (in the opinion of Newton or his contemporaries) contain an explanation of gravitational phenomena, but only a description. In both cases there came a stage at which such problems were classed as intrinsically illegitimate, and regarded exactly as the request for an explanation of why a body retains its velocity in the absence of impressed forces. While all of this may be interpreted in various ways (such as through Kuhn's theory of paradigms) the important fact for the theory of explanation is that not everything in a theory's domain is a legitimate topic for why-questions; and that what is, is not determinable *a priori*.

The second difficulty is the asymmetry revealed by the barometer, the red shift, and the flagpole examples: even if the theory implies that one condition obtains when and only when another does, it may be that it explains the one in terms of the other and not vice versa. An example which combines both the first and second difficulties is this: according to atomic physics, each chemical element has a characteristic atomic structure and a characteristic spectrum (of light emitted upon excitation). Yet the spectrum is explained by the atomic structure, and the question why a substance has that structure does not arise at all (except in the trivial sense that the questioner may need to have the terms explained to him).

To be successful, a theory of explanation must accommodate, and account for, both rejections and asymmetries. I shall now examine some attempts to come to terms with these, and gather from them the clues to the correct account.

§2.5 *Causality: the* Conditio Sine Qua Non

Why are there no longer any Tasmanian natives? Why are the Plains Indians now living on reservations? Of course it is possible to cite relevant statistics: in many areas of the world, during many periods of history, upon the invasion by a technologically advanced people, the natives were displaced and weakened culturally, physically, and economically. But such a response will not satisfy: what we want is the story behind the event.

In Tasmania, attempts to round up and contain the natives were unsuccessful, so the white settlers simply started shooting them, man, woman, and child, until eventually there were none left. On the American Plains, the whites systematically destroyed the great buffalo herds on which the Indians relied for food and clothing, thus dooming them to starvation or surrender. There you see the story, it moves by its own internal necessity, and it explains why.

I use the word 'necessity' advisedly, for that is the term that links stories and causation. According to Aristotle's *Poetics*, the right way to write a story is to construct a situation which, after the initial parameters are fixed, moves toward its conclusion with a sort of necessity, inexorably—in retrospect, 'it had to end that way'. This was to begin also the hallmark of a causal explanation. Both in literature and in science we now accept such accounts as showing only how the events could have come about in the way they did. But it may be held that, to be an explanation, a scientific account must still tell a story of how things did happen and how the events hang together, so to say.

The idea of causality in modern philosophy is that of a relation among events. Hence it cannot be identified even with efficient causation, its nearest Aristotelian relative. In the modern sense we cannot say, correctly and non-elliptically, that the salt, or the moisture in the air, caused the rusting of the knife. Instead, we must say that certain events caused the rusting: such events as dropping of the salt on the knife, the air moistening that salt, and so on. The exact phrasing is not important; that the *relata* are events (including processes and momentary or prolonged states of affairs) is very important.

But what exactly is that causal relation? Everyone will recognize Hume's question here, and recall his rejection of certain metaphysical accounts. But we do after all talk this way, we say that the knife rusted because I dropped salt on it—and, as philosophers, we must make sense of explanation. In this and the next subsection I shall discuss some attempts to explicate the modern causal relation.

When something is cited as a cause, it is not implied that it was sufficient to produce (guarantee the occurrence) of the event. I say that this plant died because it was sprayed with defoliant, while knowing that the defoliant is only 90 per cent effective. Hence, the tradition that identifies the cause as the *conditio sine qua non*: had the plant not been sprayed, it would not have died.[19]

There are two problems with restating this as: a cause is a necessary condition. In the first place, not every necessary condition is a cause; and secondly, in some straightforward sense, a cause may not be necessary, namely, alternative causes could have led to the same result. An example for the first problem is this: the existence of the knife is a necessary condition for its rusting, and the growth of the plant for its dying. But neither of these could be cited as a cause. As to the second, it is clear that the plant could have died some other way, say if I had carefully covered it totally with anti-rust paint.

J. L. Mackie proposed the definition: a cause is an insufficient but necessary part of an unnecessary but sufficient condition.[20] That sufficient condition must precede the event to be explained, of course; it must not be something like the (growth-plus death-plus rotting) of the plant if we wish to cite a cause for its death. But the first problem still stands anyway, since the existence of knife is a necessary part of the total set of conditions that led to its rusting. More worrisome is the fact that there may be no sufficient preceding conditions at all: the presence of the radium is what caused the Geiger counter to click, but atomic physics allows a non-zero probability for the counter not clicking at all under the circumstances.

For this reason (the non-availability of sufficient conditions in certain cases), Mackie's definition does not defuse the second problem either.

David Lewis has given an account in terms of counterfactual conditionals.[21] He simply equates 'A caused B' with 'if A had not happened, B would not have happened'. But it is important to understand this conditional sentence correctly, and not to think of it (as earlier logicians did) as stating that A was a necessary condition for the occurrence of B. Indeed, the 'if . . . then' is not correctly identified with any of the sorts of implication traditionally discussed in logical theory, for those obey the law of *Weakening*:

1. If A then B
   hence
   if A and C then B.

But our conditionals, in natural language, typically do not obey that law:

2. If the match is struck it will light
   hence (?)

> if the match is dunked in coffee and
> struck, it will light;

the reader will think of many other examples. The explanation of
why that 'law' does not hold is that our conditionals carry a tacit
ceteris paribus clause:

3. If the plant had not been sprayed
   (*and all else had been the same*)
   then it would not have died.

The logical effect of this tacit clause is to make the 'law' of Weakening inapplicable.

Of course, it is impossible to spell out the exact content of *ceteris
paribus*, as Goodman found in his classic discussion, for that content
changes from context to context.[22] To this point I shall have to
return. Under the circumstances, it is at least logically tenable to
say, as David Lewis does, that whenever '*A* is the (a) cause of (or:
caused) *B*' is true, it is also true that if *A* had not happened, neither
would *B* have.

But do we have a sufficient criterion here? Suppose David's alarm
clock goes off at seven a.m. and he wakes up. Now, we cite the alarm
as the cause of the awakening, and may grant, if only for the sake
of argument, that if the alarm had not sounded, he would not (then)
have woken up. But it is also true that if he had not gone to sleep
the night before, he would not have woken in the morning. This does
not seem sufficient reason to say that he woke up because he had
gone to sleep.

The response to this and similar examples is that the counterfactuals single out all the nodes in the causal net on lines leading to
the event (the awakening), whereas 'because' points to specific factors that, for one reason or other, seem especially relevant (*salient*)
in the context of our discussion. No one will deny that his going
to sleep was one of the events that 'led up' to his awakening, that
is, in the relevant part of the causal net. That part of the causal story
is objective, and which specific item is singled out for special attention depends on the context—*every* theory of causation must say
this.

Fair enough. That much context-dependence everyone will have
to allow. But I think that much more context-dependence enters this
theory through the truth-conditions of the counterfactuals themselves. So much, in fact, that we must conclude that there is nothing

in science itself—nothing in the objective description of nature that science purports to give us—that corresponds to these counterfactual conditionals.

Consider again statement (3) about the plant sprayed with defoliant. It is true in a given situation exactly if the 'all else' that is kept 'fixed' is such as to rule out death of the plant for other reasons. But who keeps what fixed? The speaker, in his mind. There is therefore a contextual variable—determining the content of that tacit *ceteris paribus* clause—which is crucial to the truth-value of the conditional statement. Let us suppose that I say to myself, *sotto voce*, that a certain fuse leads into a barrel of gunpowder, and then say out loud, 'If Tom lit that fuse there would be an explosion.' Suppose that before I came in, you had observed to yourself that Tom is very cautious, and would not light any fuse before disconnecting it, and said out loud, 'If Tom lit that fuse, there would be no explosion.' Have we contradicted each other? Is there an objective right or wrong about keeping one thing rather than another firmly in mind when uttering the antecedent 'If Tom lit that fuse . . .'? It seems rather that the proposition expressed by the sentence depends on a context, in which 'everything else being equal' takes on a definite content.

Robert Stalnaker and David Lewis give truth-conditions for conditionals using the notion of similarity among possible worlds. Thus, on one such account, 'if A then B' is true in world w exactly if B is true in the most similar world to w in which A is true. But there are many similarity relations among any set of things. Examples of the sort I have just given have long since elicited the agreement that the relevant similarity relation changes from context to context. Indeed, without that agreement, the logics of conditionals in the literature are violated by these examples.

One such example is very old: Lewis Carroll's puzzle of the three barbers. It occurs in *The Philosophy of Mr. B*rtr*nd R*ss*ll* as follows:

Allen, Brown, and Carr keep a barber's shop together; so that one of them must be in during working hours. Allen has lately had an illness of such a nature that, if Allen is out, Brown must be accompanying him. Further, if Carr is out, then, if Allen is out, Brown must be in for obvious business reasons.[23]

The above story gives rise to two conditionals, if we first suppose that Carr is out:

1. If Allen is out then Brown is out
2. If Allen is out then Brown is in

the first warranted by the remarks about Allen's illness, the second by the obvious business reasons. Lewis Carroll, thinking that 1 and 2 contradict each other, took this as a *reductio ad absurdum* of the supposition that Carr is out. R*ss*ll, construing 'if *A* then *B*' as the material conditional ('either *B* or not *A*') asserts that 1 and 2 are both true if Allen is not out, and so says that we have here only a proof that if Carr is out, then Allen is in. ('The odd part of this conclusion is that it is the one which common-sense would have drawn', he adds.)

We have many other reasons, however, for not believing the conditional of natural language to be the material conditional. In modal logic, the strict conditional is such that 1 and 2 imply that it is not possible that Allen is out. So the argument would demonstrate 'If Carr is out then it is not possible that Allen is out.' This is false; if it looks true, it does so because it is easily confused with 'It is not possible that Carr is out and Allen is out.' If we know that Carr is out we can conclude that it is false that Allen is out, not that it is impossible.

The standard logics of counterfactual conditionals give exactly the same conclusion as the modal logic of strict conditionals. However, by noting the context-dependence of these statements, we can solve the problem correctly. Statement 1 is true in a context in which we disregard business requirements and keep fixed the fact of Allen's illness; statement 2 is true if we reverse what is fixed and what is variable. Now, there can exist contexts *c* and *c'* in which 1 and 2 are true respectively, only if their common antecedent is false; thus, like R*ss*ll, we are led to the conclusion drawn by common sense.

Any of the examples, and any general form of semantics for conditionals, will lend themselves to make the same point. What sort of situation, among all the possible unrealized ones, is more like ours in the fuse example: one in which nothing new is done except that the fuse is lit, or one in which the fuse is lit after being disconnected? It all depends—similar in what respect? Similar in that no fuse is disconnected or similar in that no one is being irresponsible? Quine brought out this feature of counterfactuals—to serve another purpose—when he asked whether, if Verdi and Bizet had been compatriots, would they have been French or Italian? Finally, even if someone feels very clear on what facts should be kept fixed in the evaluation of a counterfactual conditional, he will soon realize that it is not merely the facts but the description of the facts—or, if you like, facts identified by non-extensional criteria—that matter: Danny is a man, Danny is very much interested in women, i.e. (?) in the opposite sex—if he had been a woman would he have been very much interested in men, or a Lesbian?

These puzzles cause us no difficulty, if we say that the content of 'all else being equal' is fixed not only by the sentence and the factual situation, but also by contextual factors. In that case, however, the hope that the study of counterfactuals might elucidate science is quite mistaken: scientific propositions are not context-dependent in any essential way, so if counterfactual conditionals are, then science neither contains nor implies counterfactuals.

The truth-value of a conditional depends in part on the context. Science does not imply that the context is one way or another. There-fore science does not imply the truth of any counterfactual—except in the limiting case of a conditional with the same truth-value in all contexts. (Such limiting cases are ones in which the scientific theory plus the antecedent strictly imply the consequent, and for them logical laws like Weakening and Contraposition are valid, so that they are useless for the application to explanation which we are at present exploring.)

There was at one point a hope, expressed by Goodman, Reichen-bach, Hempel, and others, that counterfactual conditionals provide an objective criterion for what is a law of nature, or at least, a lawlike statement (where a law is a true lawlike statement). A merely general truth was to be distinguished from a law because the latter, and not the former, implies counterfactuals. This idea must be inverted: if laws imply counterfactuals then, because counterfactuals are con-text-dependent, the concept of law does not point to any objective distinction in nature.

If, as I am inclined to agree, counterfactual language is proper to explanation, we should conclude that explanation harbours a sig-nificant degree of context-dependence.

§2.6 *Causality: Salmon's Theory*

The preceding subsection began by relating causation to stories, but the accounts of causality it examined concentrated on the links between particular events. The problems that appeared may therefore have resulted from the concentration on 'local properties' of the story. An account of causal explanation which focuses on extended processes has recently been given by Wesley Salmon.[24]

In his earlier theory, to the effect that an explanation consists in listing statistically relevant factors, Salmon had asked 'What more could one ask of an explanation?' He now answers this question:

What does explanation offer, over and above the inferential capacity of prediction and retrodiction ...? It provides knowledge of the mechanisms of *production* and *propagation* of structure in the world. That goes some distance beyond mere recognition of regularities, and of the possibility of subsuming particular phenomena thereunder.[25]

The question, what is the causal relation? is now replaced by: what is a causal process? and, what is a causal interaction? In his answer to these questions, Salmon relies to a large extent on Reichenbach's theory of the common cause, which we encountered before. But Salmon modifies this theory considerably.

A process is a spatio-temporally continuous series of events. The continuity is important, and Salmon blames some of Hume's difficulties on his picture of processes as chains of events with discrete links.[26] Some processes are causal, or genuine processes, and some are pseudo-processes. For example, if a car moves along a road, its shadow moves along that road too. The series of events in which the car occupies successive points on that road is a genuine causal process. But the movement of the shadow is merely a pseudo-process, because, intuitively speaking, the position of the shadow at later times is not caused by its position at earlier times. Rather, there is shadow *here* now because there is a car here now, and not because there was shadow *there* then.

Reichenbach tried to give a criterion for this distinction by means of probabilistic relations.[27] The series of events A_r is a causal process provided

(1) the probability of A_{r+s} given A_r is greater than or equal to the probability of A_{r+s} given A_{r-t}, which is in turn greater than the probability of A_{r+s} *simpliciter*.

This condition does not yet rule out pseudo-processes, so we add that each event in the series *screens off* the earlier ones from the later ones:

(2) the probability of A_{r+s} given both A_r and A_{r-t} is just that of A_{r+s} given A_r

and, *in addition*, there is no other series of events B_r which screens off A_{r+s} from A_r for all r. The idea in the example is that if A_{r+s} is the position of the shadow at time $r+s$, then B_r is the position of the car at time $r+s$.

This is not satisfactory for two reasons. The first is that (1) reminds one of a well-known property of stochastic processes, called the Markov property, and seems to be too strong to go into the definition of causal processes. Why should not the whole history of the process up to time r give more information about what happens later than the state at time r does by itself? The second problem is that in the addition to (2) we should surely add that B_r must itself be a genuine causal process? For otherwise the movement of the car is not a causal process either, since the movement of the shadow will screen off successive positions of the car from each other. But if we say that B_r must be a genuine process in this stipulation, we have landed in a regress.

Reichenbach suggested a second criterion, called the *mark method* and (presumably because it stops the threatened regress) Salmon prefers that.

If a fender is scraped as a result of a collision with a stone wall, the mark of that collision will be carried on by the car long after the interaction with the wall occurred. The shadow of a car moving along the shoulder is a pseudo-process. If it is deformed as it encounters a stone wall, it will immediately resume its former shape as soon as it passes by the wall. It will not transmit a mark or modification.[28]

So if the process is genuine then interference with an earlier event will have effects on later events in that process. However, thus phrased, this statement is blatantly a causal claim. How shall we explicate 'interference' and 'effects'? Salmon will shortly give an account of causal interactions (see below) but begins by appealing to his 'at-at' theory of motion. The movement of the car consists simply in being *at* all these positions *at* those various times. Similarly, the propagation of the mark consists simply in the mark being there, in those later events. There is not, over and above this, a special propagation relation.

However, there is more serious cause for worry. We cannot define a genuine process as one that *does* propagate a mark in this sense. There are features which the shadow carries along in that 'at-at' sense, such as that its shape is related, at all times, in a certain topologically definable way to the shape of the car, and that it is black.

Other special marks are not always carried—imagine part of a rocket's journey during which it encounters nothing else. So what we need to say is that the process is genuine if, *were* there to be a given sort of interaction at an early stage, there *would be* certain marks in the later stages. At this point, I must refer back to the preceding section for a discussion of such counterfactual assertions.

We can, at this point, relativize the notions used to the theory accepted. About some processes, our theory *implies* that certain interactions at an early stage will be followed by certain marks at later stages. Hence we can say that, *relative to the theory* certain processes are classifiable as genuine and others as pseudo-processes. What this does not warrant is regarding the distinction as an objective one. However, if the distinction is introduced to play a role in the theory of explanation, and if explanation is a relation of theory to fact, this conclusion does not seem to me a variation on Salmon's theory that would defeat its purpose.[29]

Turning now to causal interactions, Salmon describes two sorts. These interactions are the 'nodes' in the causal net, the 'knots' that combine all those causal processes into a causal structure. Instead of 'node' or 'knot' Reichenbach and Salmon also use 'fork' (as in 'the road forks'). Reichenbach described one sort, the *conjunctive fork* which occurs when an event C, belonging to two processes, is the *common cause* of events A and B, in those separate processes, occurring after C. Here common cause is meant in Reichenbach's original sense:

(3) $P(A \& B/C) = P(A/C) . P(B/C)$
(4) $P(A \& B/\bar{C}) = P(A/\bar{C}) . P(B/\bar{C})$
(5) $P(A/C) > P(A/\bar{C})$
(6) $P(B/C) > P(B/\bar{C})$

which, as noted in Chapter 2, entails that there is a positive correlation between A and B.

In order to accommodate the recalcitrant examples (see Chapter 2) Salmon introduced in addition the *interactive fork*, which is like the preceding one except that (3) is changed to

(3*) $P(A \& B/C) > P(A/C) . P(B/C)$

These forks then combine the genuine causal processes, once identified, into the causal net that constitutes the natural order.

Explanation, on Salmon's new account, consists therefore in exhibiting the relevant part of the causal net that leads up to the events that are to be explained. In some cases we need only point to a single causal process that leads up to the event in question. In other cases we are asked to explain the confluence of events, or a positive correlation, and we do so by tracing them back to forks, that is, common origins of the processes that led up to them.

Various standard problems are handled. The sequence, barometer falling–storm coming, is not a causal process since the relevance of the first to the second is screened off by the common cause of atmospheric conditions. When paresis is explained by mentioning latent untreated syphilis, one is clearly pointing to the causal process, whatever it is, that leads from one to the other—or to their common cause, whatever that is. It must of course be a crucial feature of this theory that ordinary explanations are 'pointers to' causal processes and interactions which would, if known or described in detail, give the full explanation.

If that is correct, then each explanation must have, as cash-value, some tracing back (which is possible in principle) of separate causal processes to the forks that connect them. There are various difficulties with this view. The first is that to be a causal process, the sequence of events must correspond to a continuous spatio-temporal trajectory. In quantum mechanics, this requirement is not met. It was exactly the crucial innovation in the transition from the Bohr atom of 1913 to the new quantum theory of 1924, that the exactly defined orbits of the electrons were discarded. Salmon mentions explicitly the limitation of this account to macroscopic phenomena (though he does discuss Compton scattering). This limitation is serious, for we have no independent reason to think that explanation in quantum mechanics is essentially different from elsewhere.

Secondly, many scientific explanations certainly do not look as if they are causal explanations in Salmon's sense. A causal law is presumably one that governs the temporal development of some process or interaction. There are also 'laws of coexistence', which give limits to possible states or simultaneous configurations. A simple example is Boyle's law for gases (temperature is proportional to volume times pressure, at any given time); another, Newton's law of gravitation; another, Pauli's exclusion principle. In some of these cases we can say that they (or their improved counterparts) were later deduced from theories that replaced 'action at a distance' (which is not action at all, but a constraint on simultaneous states) with 'action by contact'. But suppose they were not so replaceable—would that mean that they could not be used in genuine explanations?

Salmon himself gives an example of explanation 'by common cause' which actually does not seem to fit his account. By observations on Brownian motion, scientists determined Avogadro's number, that is, the number of molecules in one mole of gas. By quite different observations, on the process of electrolysis, they determined the number of electron charges equal to one Faraday, that is, to the amount of electric charge needed to deposit one mole of a monovalent metal. These two numbers are equal. On the face of it, this equality is astonishing; but physics can explain this equality by deducing it from the basic theories governing both sorts of phenomena. The common cause Salmon identifies here is the basic mechanism—atomic and molecular structure—postulated to account for these phenomena. But surely it is clear that, however much the adduced explanation may deserve the name 'common cause', it does not point to a relationship between events (in Brownian motion on specific occasions and in electrolysis on specific occasions) which is traced back via causal processes to forks connecting these processes. The explanation is rather that the number found in experiment A at time t is the same as that found in totally independent experiment B at *any* other time t', because of the *similarity* in the physically independent causal processes observed on those two different occasions.

Many highly theoretical explanations at least look as if they escape Salmon's account. Examples here are explanations based on principles of least action, based on symmetry considerations, or, in relativistic theories, on information that relates to space–time as a whole, such as specification of the metric or gravitational field.

The conclusion suggested by all this is that the type of explanation characterized by Salmon, though apparently of central importance, is still at most a subspecies of explanations in general.

§2.7 *The Clues of Causality*

Let us agree that science gives us a picture of the world as a net of interconnected events, related to each other in a complex but orderly way. The difficulties we found in the preceding two sections throw some doubt on the adequacy of the terminology of cause and causality to describe that picture; but let us not press this doubt further. The account of explanation suggested by the theories examined can now be restated in general terms as follows:

(1) Events are enmeshed in a net of causal relations
(2) What science describes is that causal net
(3) Explanation of why an event happens consists (typically) in an exhibition of salient factors in the part of the causal net formed by lines 'leading up to' that event
(4) Those salient factors mentioned in an explanation constitute (what are ordinarily called) the *cause(s)* of that event.

There are two clear reasons why, when the topic of explanation comes up, attention is switched from the causal net as a whole (or even the part that converges on the event in question) to 'salient factors'. The first reason is that any account of explanation must make sense of common examples of explanation—especially cases typically cited as scientific explanations. In such actual cases, the reasons cited are particular prior events or initial conditions or combinations thereof. The second reason is that no account of explanation should imply that we can never give an explanation—and to describe the whole causal net in any connected region, however small, is in almost every case impossible. So the least concession one would have to make is to say that the explanation need say no more than that *there is* a structure of causal relations of *a certain sort*, which could *in principle* be described in detail: the salient features are what picks out the 'certain sort'.

Interest in causation as such focuses attention on (1) and (2), but interest in explanation requires us to concentrate on (3) and (4). Indeed, from the latter point of view, it is sufficient to guarantee the truth of (1) and (2) by *defining*

the causal net = whatever structure of relations science describes

and leaving to those interested in causation as such the problem of describing that structure in abstract but illuminating ways, if they wish.

Could it be that the explanation of a fact or event nevertheless resides solely in that causal net, and that *any* way of drawing attention to it explains? The answer is *no*; in the case of causal explanation, the *explanation* consists in drawing attention to certain ('special', 'important') features of the causal net. Suppose for example that I wish to explain the extinction of the Irish elk. There is a very large class of factors that preceded this extinction and was statistically relevant to it—even very small increases in speed, contact area of the hoof, height, distribution of weight in the body, distribution of food supply, migration habits, surrounding fauna and flora—we know from selection theory that under proper conditions any variation in these can be decisive in the survival of the species. But although, if some of these had been different, the Irish elk would have survived, they are not said to provide the explanation of why it is now extinct. The explanation given is that the process of sexual selection favoured males with large antlers, and that these antlers were, in the environment where they lived, encumbering and the very opposite of survival-adaptive. The other factors I mentioned are not spurious causes, or screened off by the development of these huge and cumbersome antlers, because the extinction was the total effect of many contributing factors; but those other factors are not the salient factors.

We turn then to those salient features that are cited in explanation—those referred to as 'the cause(s)' or 'the real cause(s)'. Various philosophical writers, seeking for an objective account of explanation, have attempted to state criteria that single out those special factors. I shall not discuss their attempts. Let me just cite a small survey of their answers: Lewis White Beck says that the cause is that factor over which we have most control; Nagel argues that it is often exactly that factor which is not under our control; Braithwaite takes the salient factors to be the unknown ones; and David Bohm takes them to be the factors which are the most variable.[30]

Why should different writers have given such different answers? The reason was exhibited, I think, by Norwood Russell Hanson, in his discussion of causation.

There are as many causes of x as there are explanations of x. Consider how the cause of death might have been set out by a physician as 'multiple haemorrhage', by the barrister as 'negligence on the part of the driver', by a carriage-builder as 'a defect in the brakeblock construction', by a civic planner as 'the presence of tall shrubbery at that turning'.[31]

In other words, the salient feature picked out as 'the cause' in that complex process, is salient to a given person because of his orientation, his interests, and various other peculiarities in the way he approaches or comes to know the problem—contextual factors.

It is important to notice that in a certain sense these different answers cannot be combined. The civic planner 'keeps fixed' the mechanical constitution of the car, and gives his answer in the conviction that regardless of the mechanical defects, which made a fast stop impossible, the accident need not have happened. The mechanic 'keeps fixed' the physical environment; despite the shrubbery obscuring vision, the accident need not have happened if the brakes had been better. What the one varies, the other keeps fixed, and you cannot do both at once. In other words, the selection of the salient causal factor is not simply a matter of pointing to the most interesting one, not like the selection of a tourist attraction; it is a matter of *competing* counterfactuals.

We must accordingly agree with the Dutch philosopher P. J. Zwart who concludes, after examining the above philosophical theories,

It is therefore not the case that the meaning of the sentence 'A is the cause of B' depends on the nature of the phenomena A and B, but that this meaning depends on the context in which this sentence is uttered. The nature of A and B will in most cases also play a role, indirectly, but it is in the first place the orientation or the chosen point of view of the speaker that determines what the word cause is used to signify.[32]

In conclusion, then, this look at accounts of causation seems to establish that explanatory factors are to be chosen from a range of factors which are (or which the scientific theory lists as) objectively relevant in certain special ways—but that the choice is then determined by other factors that vary with the context of the explanation request. To sum up: no factor is explanatorily relevant unless it is scientifically relevant; and among the scientifically relevant factors, context determines explanatorily relevant ones.

§2.8 *Why-questions*

Another approach to explanation was initiated by Sylvain Bromberger in his study of why-questions.[33] After all, a why-question is a request for explanation. Consider the question:

1. Why did the conductor become warped during the short circuit?

This has the general form

2. Why (is it the case that) *P*?

where *P* is a statement. So we can think of 'Why' as a function that turns statements into questions.

Question 1 *arises*, or *is in order*, only if the conductor did indeed become warped then. If that is not so, we do not try to answer the question, but say something like: 'You are under a false impression, the conductor became warped much earlier,' or whatever. Hence Bromberger calls the statement that *P* the *presupposition* of the question *Why P?* One form of the rejection of explanation requests is clearly the denial of the presupposition of the corresponding why-question.

I will not discuss Bromberger's theory further here, but turn instead to a criticism of it. The following point about why-questions has been made in recent literature by Alan Garfinkel and Jon Dorling, but I think it was first made, and discussed in detail, in unpublished work by Bengt Hannson circulated in 1974.[34] Consider the question

3. Why did Adam eat the apple?

This same question can be construed in various ways, as is shown by the variants:

3a. Why was it Adam who ate the apple?
3b. Why was it the apple Adam ate?
3c. Why did Adam *eat* the apple?

In each case, the canonical form prescribed by Bromberger (as in 2 above) would be the same, namely

4. Why (is it the case that) (Adam ate the apple)?

yet there are three different explanation requests here.

The difference between these various requests is that they point to different contrasting alternatives. For example, 3b may ask why Adam ate *the apple* rather than some other fruit in the garden, while 3c asks perhaps why Adam *ate* the apple rather than give it back to Eve untouched. So to 3b, 'because he was hungry' is not a good answer, whereas to 3c it is. The correct general, underlying structure of a why-question is therefore

5. Why (is it the case that) *P in contrast to* (other members of) *X*?

where *X*, the *contrast-class*, is a set of alternatives. *P* may belong to *X* or not; further examples are:

Why did the sample burn green (rather than some other colour)?
Why did the water and copper reach equilibrium temperature 22.5°C (rather than some other temperature)?

In these cases the contrast-classes (colours, temperatures) are 'obvious'. In general, the contrast-class is not explicitly described because, *in context*, it is clear to all discussants what the intended alternatives are.

This observation explains the tension we feel in the paresis example. If a mother asks why her eldest son, a pillar of the community, mayor of his town, and best beloved of all her sons, has this dread disease, we answer: because he had latent untreated syphilis. But if that question is asked about this same person, immediately after a discussion of the fact that everyone in his country club has a history of untreated syphilis, *there is no answer*. The reason for the difference is that in the first case the contrast-class is the mother's sons, and in the second, the members of the country club, contracting paresis. Clearly, an answer to a question of form 5 must adduce information that *favours P in contrast to* other members of X. Sometimes the availability of such information depends strongly on the choice of X.

These reflections have great intuitive force. The distinction made is clearly crucial to the paresis example and explains the sense of ambiguity and tension felt in earlier discussion of such examples. It also gives us the right way to explicate such assertions as: individual events are never explained, we only explain a particular event *qua* event of a certain kind. (We can explain *this* decay of a uranium atom *qua* decay of a uranium atom, but not *qua* decay of a uranium atom at *this* time.)

But the explication of what it is for an answer to favour one alternative over another proves difficult. Hannson proposed: answer A is a good answer to (Why P in contrast to X?) exactly if the probability of P given A is higher than the average probability of members of X given A. But this proposal runs into most of the old difficulties. Recall Salmon's examples of irrelevancy: the probability of recovery from a cold *given* administration of vitamin C is nearly one, while the probability of not recovering *given* the vitamins is nearly zero. So by Hannson's criterion it would be a good answer—even if taking vitamin C has no effect on recovery from colds one way or the other.

Also, the asymmetries are as worrisome as ever. By Hannson's criterion, the length of the shadow automatically provides a good explanation of the height of the flagpole. And 'because the barometer fell' is a good answer to 'why is there a storm?' (upon selection of the 'obvious' contrast-classes, of course). Thus it seems that reflection on the contrast-class serves to solve some of our problems, but not all.

§2.9 *The Clues Elaborated*

The discussions of causality and of why-questions seem to me to provide essential clues to the correct account of explanation. In the former we found that an explanation often consists in listing salient factors, which point to a complete story of how the event happened. The effect of this is to eliminate various alternative hypotheses about how this event did come about, and/or eliminate puzzlement concerning how the event could have come about. But salience is context-dependent, and the selection of the correct 'important' factor depends on the range of alternatives contemplated in that context. In N. R. Hanson's example, the barrister wants this sort of weeding out of hypotheses about the death relevant to the question of legal accountability; the carriage-builder, a weeding out of hypotheses about structural defects or structural limitations under various sorts of strain. *The context,* in other words, *determines relevance* in a way that goes well beyond the statistical relevance about which our scientific theories give information.

This might not be important if we were not concerned to find out exactly how having an explanation goes beyond merely having an acceptable theory about the domain of phenomena in question. But that is exactly the topic of our concern.

In the discussion of why-questions, we have discovered a further contextually determined factor. The range of hypotheses about the event which the explanation must 'weed out' or 'cut down' is not determined solely by the interests of the discussants (legal, mechanical, medical) but also by a range of contrasting alternatives to the event. This *contrast-class* is also determined by context.

It might be thought that when we request a *scientific* explanation, the relevance of possible hypotheses, and also the contrast-class are automatically determined. But this is not so, for both the physician and the motor mechanic are asked for a scientific explanation. The physician explains the fatality *qua* death of a human organism, and the mechanic explains it *qua* automobile crash fatality. To ask that their explanations be scientific is only to demand that they rely on scientific theories and experimentation, not on old wives' tales. Since any explanation of an individual event must be an explanation of that event *qua* instance of a certain kind of event, nothing more can be asked.

The two clues must be put together. The description of some account as an explanation of a given fact or event, is incomplete. It can only be an explanation with respect to a certain *relevance relation* and a certain *contrast-class*. These are contextual factors, in that they are determined neither by the totality of accepted scientific theories, nor by the event or fact for which an explanation is requested. It is sometimes said that an Omniscient Being would have a complete explanation, whereas these contextual factors only bespeak our limitations due to which we can only grasp one part or aspect of the complete explanation at any given time. But this is a mistake. If the Omniscient Being has no specific interests (legal, medical, economic; or just an interest in optics or thermodynamics rather than chemistry) and does not abstract (so that he never thinks of Caesar's death *qua* multiple stabbing, or *qua* assassination), then no why-questions ever arise for him in any way at all—and he does not have any explanation in the sense that we have explanations. If he does have interests, and does abstract from individual peculiarities in his thinking about the world, then his why-questions are as essentially context-dependent as ours. In either case, his advantage is that he always has all the information needed to answer any specific explanation request. But that information is, in and by itself, not an explanation; just as a person cannot be said to be older, or a neighbour, except in relation to others.

§3. *Asymmetries of Explanation: A Short Story*

§3.1 *Asymmetry and Context: the Aristotelian Sieve*

That vexing problem about paresis, where we seem both to have and not to have an explanation, was solved by reflection on the contextually supplied contrast-class. The equally vexing, and much older, problem of the asymmetries of explanation, is illuminated by reflection on the other main contextual factor: contextual relevance.

If that is correct, if the asymmetries of explanation result from a contextually determined relation of relevance, then it must be the case that these asymmetries can at least sometimes be reversed by a change in context. In addition, it should then also be possible to account for specific asymmetries in terms of the interests of questioner and audience that determine this relevance. These considerations provide a crucial test for the account of explanation which I propose.

Fortunately, there is a precedent for this sort of account of the asymmetries, namely in Aristotle's theory of science. It is traditional to understand this part of his theory in relation to his metaphysics; but I maintain that the central aspects of his solution to the problem of asymmetry of explanations are independently usable.[35]

Aristotle gave examples of this problem in the *Posterior Analytics* I, 13; and he developed a typology of explanatory factors ('the four causes'). The solution is then simply this. Suppose there are a definite (e.g. four) number of types of explanatory factors (i.e. of relevance relations for why-questions). Suppose also that relative to our background information and accepted theories, the propositions *A* and *B* are equivalent. It may then still be that these two propositions describe factors of different types. Suppose that in a certain context, our interest is in the mode of production of an event, and 'Because *B*' is an acceptable answer to 'Why *A*?'. Then it may well be that *A* does not describe any mode of production of anything, so that, *in this same context*, 'Because *A*' would not be an acceptable answer to 'Why *B*?'.

Aristotle's lantern example (*Posterior Analytics* II, 11) shows that he recognized that in different contexts, verbally the same why-question may be a request for different types of explanatory factors. In modern dress the example would run as follows. Suppose a father asks his teenage son, 'Why is the porch light on?' and the son replies 'The porch switch is closed and the electricity is reaching the bulb through that switch.' At this point you are most likely to feel that the son is being impudent. This is because you are most likely to think that the sort of answer the father needed was something like: 'Because we are expecting company.' But it is easy to imagine a less likely question context: the father and son are re-wiring the house and the father, unexpectedly seeing the porch light on, fears that he has caused a short circuit that bypasses the porch light switch. In the second case, he is *not* interested in the human expectations or desires that led to the depressing of the switch.

Aristotle's fourfold typology of causes is probably an over-simplification of the variety of interests that can determine the selection of a range of relevant factors for a why-question. But in my opinion, appeal to some such typology will successfully illuminate the asymmetries (and also the rejections, since no factor of a *particular* type may lead to a telling answer to the why-question). If that is so then, as I said before, asymmetries must be at least sometimes reversible through a change in context. The story which follows is meant to illustrate this. As in the lantern (or porch light) example, the relevance changes from one sort of efficient cause to another, the second being a person's desires. As in all explanations, the correct answer consists in the exhibition of a single factor in the causal net, which is made salient in that context by factors not overtly appearing in the words of the question.

§3.2 *'The Tower and the Shadow'*

During my travels along the Saône and Rhône last year, I spent a day and night at the ancestral home of the Chevalier de St. X ..., an old friend of my father's. The Chevalier had in fact been the French liaison officer attached to my father's brigade in the first war, which had—if their reminiscences are to be trusted—played a not insignificant part in the battles of the Somme and Marne.

The old gentleman always had *thé à l'Anglaise* on the terrace at five o'clock in the evening, he told me. It was at this meal that a strange incident occurred; though its ramifications were of course not yet perceptible when I heard the Chevalier give his simple explanation of the length of the shadow which encroached upon us there on the terrace. I had just eaten my fifth piece of bread and butter and had begun my third cup of tea when I chanced to look up. In the dying light of that late afternoon, his profile was sharply etched against the granite background of the wall behind him, the great aquiline nose thrust forward and his eyes fixed on some point behind my left shoulder. Not understanding the situation at first, I must admit that to begin with, I was merely fascinated by the sight of that great hooked nose, recalling my father's claim that this had once served as an effective weapon in close combat with a German grenadier. But I was roused from this brown study by the Chevalier's voice.

'The shadow of the tower will soon reach us, and the terrace will turn chilly. I suggest we finish our tea and go inside.'

I looked around, and the shadow of the rather curious tower I had earlier noticed in the grounds, had indeed approached to within a yard from my chair. The news rather displeased me, for it was a fine evening; I wished to remonstrate but did not well know how, without overstepping the bounds of hospitality. I exclaimed,

'Why must that tower have such a long shadow? This terrace is so pleasant!'

His eyes turned to rest on me. My question had been rhetorical, but he did not take it so.

'As you may already know, one of my ancestors mounted the scaffold with Louis XVI and Marie Antoinette. I had that tower erected in 1930 to mark the exact spot where it is said that he greeted the Queen when she first visited this house, and presented her with a peacock made of soap, then a rare substance. Since the Queen would have been one hundred and seventy-five years old in 1930, had she lived, I had the tower made exactly that many feet high.'

It took me a moment to see the relevance of all this. Never quick at sums, I was at first merely puzzled as to why the measurement should have been in feet; but of course I already knew him for an Anglophile. He added drily, 'The sun not being alterable in its course, light travelling in straight lines, and the laws of trigonometry being immutable, you will perceive that the length of the shadow is determined by the height of the tower.' We rose and went inside.

I was still reading at eleven that evening when there was a knock at my door. Opening it I found the housemaid, dressed in a somewhat old-fashioned black dress and white cap, whom I had perceived hovering in the background on several occasions that day. Courtseying prettily, she asked, 'Would the gentleman like to have his bed turned down for the night?'

I stepped aside, not wishing to refuse, but remarked that it was very late—was she kept on duty to such hours? No, indeed, she answered, as she deftly turned my bed covers, but it had occurred to her that some duties might be pleasures as well. In such and similar philosophical reflections we spent a few pleasant hours together, until eventually I mentioned casually how silly it seemed to me that the tower's shadow ruined the terrace for a prolonged, leisurely tea.

At this, her brow clouded. She sat up sharply. 'What exactly did he tell you about this?' I replied lightly, repeating the story about Marie Antoinette, which now sounded a bit far-fetched even to my credulous ears.

'The *servants* have a different account', she said with a sneer that was not at all becoming, it seemed to me, on such a young and pretty face. 'The truth is quite different, and has nothing to do with ancestors. That tower marks the spot where he killed the maid with whom he had been in love to the point of madness. And the height of the tower? He vowed that shadow would cover the terrace where he first proclaimed his love, with every setting sun—that is why the tower had to be so high.'

I took this in but slowly. It is never easy to assimilate unexpected truths about people we think we know—and I have had occasion to notice this again and again.

'Why did he kill her?' I asked finally.

'Because, sir, she dallied with an English brigadier, an overnight guest in this house.' With these words she arose, collected her bodice and cap, and faded through the wall beside the doorway.

I left early the next morning, making my excuses as well as I could.

§4. *A Model for Explanation*

I shall now propose a new theory of explanation. An explanation is not the same as a proposition, or an argument, or list of propositions; it is an *answer*. (Analogously, a son is not the same as a man, even if all sons are men, and every man is a son.) An explanation is an answer to a why-question. So, a theory of explanation must be a theory of why-questions.

To develop this theory, whose elements can all be gleaned, more or less directly, from the preceding discussion, I must first say more about some topics in formal pragmatics (which deals with context-dependence) and in the logic of questions. Both have only recently become active areas in logical research, but there is general agreement on the basic aspects to which I limit the discussion.

§4.1 *Contexts and Propositions*[36]

Logicians have been constructing a series of models of our language, of increasing complexity and sophistication. The phenomena they aim to save are the surface grammar of our assertions and the inference patterns detectable in our arguments. (The distinction between logic and theoretical linguistics is becoming vague, though logicians' interests focus on special parts of our language, and require a less faithful fit to surface grammar, their interests remaining in any case highly theoretical.) Theoretical entities introduced by logicians in their models of language (also called 'formal languages') include domains of discourse ('universes'), possible words, accessibility ('relative possibility') relations, facts and propositions, truth-values, and, lately, contexts. As might be guessed, I take it to be part of empiricism to insist that the adequacy of these models does not require all their elements to have counterparts in reality. They will be good if they fit those phenomena to be saved.

Elementary logic courses introduce one to the simplest models, the languages of sentential and quantificational logic which, being the simplest, are of course the most clearly inadequate. Most logic teachers being somewhat defensive about this, many logic students, and other philosophers, have come away with the impression that the over-simplifications make the subject useless. Others, impressed with such uses as elementary logic does have (in elucidating classical mathematics, for example), conclude that we shall not understand natural language until we have seen how it can be regimented so as to fit that simple model of horseshoes and truth tables.

In elementary logic, each sentence corresponds to exactly one pro-position, and the truth-value of that sentence depends on whether the proposition in question is true in the actual world. This is also true of such extensions of elementary logic as free logic (in which not all terms need have an actual referent), and normal modal logic (in which non-truth functional connectives appear), and indeed of almost all the logics studied until quite recently.

But, of course, sentences in natural language are typically context-dependent; that is, which proposition a given sentence expresses will vary with the context and occasion of use. This point was made early on by Strawson, and examples are many:

> 'I am happy now' is true in context x exactly if the speaker in context x is happy at the time of context x,

where a context of use is an actual occasion, which happened at a definite time and place, and in which are identified the speaker (referent of 'I'), addressee (referent of 'you'), person discussed (referent of 'he'), and so on. That contexts so conceived are idealiza-tions from real contexts is obvious, but the degree of idealization may be decreased in various ways, depending on one's purposes of study, at the cost of greater complexity in the model constructed.

What must the context specify? The answer depends on the sentence being analysed. If that sentence is

> Twenty years ago it was still possible to prevent the threatened population explosion in that country, but now it is too late

the model will contain a number of factors. First, there is a set of possible worlds, and a set of contexts, with a specification for each context of the world of which it is a part. Then there will be for each world a set of entities that exist in that world, and also various relations of relative possibility among these worlds. In addition there is time, and each context must have a time of occurrence. When we evaluate the above sentence we do so relative to a context and a world. Varying with the context will be the referents of 'that country' and 'now', and perhaps also the relative possibility relation used to interpret 'possible', since the speaker may have intended one of several senses of possibility.

This sort of interpretation of a sentence can be put in a simple general form. We first identify certain entities (mathematical constructs) called propositions, each of which has a truth-value in each possible world. Then we give the context as its main task the job of selecting, for each sentence, the proposition it expresses 'in that context'. Assume as a simplification that when a sentence contains no indexical terms (like 'I', 'that', 'here', etc.), then all contexts select the same proposition for it. This gives us an easy intuitive handle on what is going on. If A is a sentence in which no indexical terms occur, let us designate as $|A|$ the proposition which it expresses in every context. Then we can generally (though not necessarily always) identify the proposition expressed by any sentence in a given context as the proposition expressed by some indexical-free sentence. For example:

> In context x, 'Twenty years ago it was still possible to prevent the population explosion in that country' expresses the proposition 'In 1958, it is (tenseless) possible to prevent the population explosion in India'

To give another example, in the context of my present writing, 'I am here now' expresses the proposition that Bas van Fraassen is in Vancouver, in July 1978.

This approach has thrown light on some delicate conceptual issues in philosophy of language. Note for example that 'I am here' is a sentence which is true no matter what the facts are and no matter what the world is like, and no matter what context of usage we consider. Its truth is ascertainable *a priori*. But the proposition expressed, that van Fraassen is in Vancouver (or whatever else it is) is not at all a necessary one: I might not have been here. Hence, a clear distinction between *a priori* ascertainability and necessity appears.

The context will generally select the proposition expressed by a given sentence *A* via a selection of referents for the terms, extensions for the predicates, and functions for the functors (i.e. syncategorematic words like 'and' or 'most'). But intervening contextual variables may occur at any point in these selections. Among such variables there will be the assumptions taken for granted, theories accepted, world-pictures or paradigms adhered to, in that context. A simple example would be the range of conceivable worlds admitted as possible by the speaker; this variable plays a role in determining the truth-value of his modal statements in that context, relative to the 'pragmatic presuppositions'. For example, if the actual world is really the only possible world there is (which exists) then the truth-values of modal statements in that context but *tout court* will be very different from their truth-values relative to those pragmatic presuppositions—and only the latter will play a significant role in our understanding of what is being said or argued in that context.

Since such a central role is played by propositions, the family of propositions has to have a fairly complex structure. Here a simplifying hypothesis enters the fray: propositions can be uniquely identified through the worlds in which they are true. This simplifies the model considerably, for it allows us to identify a proposition with a set of possible worlds, namely, the set of worlds in which it is true. It allows the family of propositions to be a complex structure, admitting of interesting operations, while keeping the structure of each individual proposition very simple.

Such simplicity has a cost. Only if the phenomena are simple enough, will simple models fit them. And sometimes, to keep one part of a model simple, we have to complicate another part. In a number of areas in philosophical logic it has already been proposed to discard that simplifying hypothesis, and to give propositions more 'internal structure'. As will be seen below, problems in the logic of explanation provide further reasons for doing so.

§4.2 *Questions*

We must now look further into the general logic of questions. There are of course a number of approaches; I shall mainly follow that of Nuel Belnap, though without committing myself to the details of his theory.[37]

A theory of questions must needs be based on a theory of propositions, which I shall assume given. A *question* is an abstract entity;

it is expressed by an *interrogative* (a piece of language) in the same sense that a proposition is expressed by a declarative sentence. Almost anything can be an appropriate response to a question, in one situation or another; as 'Peccavi' was the reply telegraphed by a British commander in India to the question how the battle was going (he had been sent to attack the province of Sind).[38] But not every response is, properly speaking, an answer. Of course, there are degrees; and one response may be more or less of an answer than another. The first task of a theory of questions is to provide some typology of answers. As an example, consider the following question, and a series of responses:

Can you get to Victoria both by ferry and by plane?
(*a*) Yes.
(*b*) You can get to Victoria both by ferry and by plane.
(*c*) You can get to Victoria by ferry.
(*d*) You can get to Victoria both by ferry and by plane, but the ferry ride is not to be missed.
(*e*) You can certainly get to Victoria by ferry, and that is something not to be missed.

Here (*b*) is the 'purest' example of an answer: it gives enough information to answer the question completely, but no more. Hence it is called a *direct answer*. The word 'Yes' (*a*) is a *code* for this answer.

Responses (*c*) and (*d*) depart from that direct answer in opposite directions: (*c*) says properly less than (*b*)—it is implied by (*b*)—while (*d*), which implies (*b*), says more. Any proposition implied by a direct answer is called a *partial answer* and one which implies a direct answer is a *complete answer*. We must resist the temptation to say that therefore an answer, *tout court*, is any combination of a partial answer with further information, for in that case, every proposition would be an answer to any question. So let us leave (*e*) unclassified for now, while noting it is still 'more of an answer' than such responses as 'Gorilla!' (which is a response given to various questions in the film *Ich bin ein Elephant, Madam*, and hence, I suppose, still more of an answer than some). There may be some quantitative notion in the background (a measure of the extent to which a response really 'bears on' the question) or at least a much more complete typology (some more of it is given below), so it is probably better not to try and define the general term 'answer' too soon.

The basic notion so far is that of direct answer. In 1958, C. L.

Hamblin introduced the thesis that a question is uniquely identifiable through its answers.[39] This can be regarded as a simplifying hypothesis of the sort we come across for propositions, for it would allow us to identify a question with the set of its direct answers. Note that this does not preclude a good deal of complexity in the determination of exactly what question is expressed by a given interrogative. Also, the hypothesis does not identify the question with the disjunction of its direct answers. If that were done, the clearly distinct questions

Is the cat on the mat?
    direct answers: The cat is on the mat.
                    The cat is not on the mat.
Is the theory of relativity true?
    direct answers: The theory of relativity is true.
                    The theory of relativity is not true.

would be the same (identified with the tautology) if the logic of propositions adopted were classical logic. Although this simplifying hypothesis is therefore not to be rejected immediately, and has in fact guided much of the research on questions, it is still advisable to remain somewhat tentative towards it.

Meanwhile we can still use the notion of direct answer to define some basic concepts. One question Q may be said to *contain* another, Q', if Q' is answered as soon as Q is—that is, every complete answer to Q is also a complete answer to Q'. A question is *empty* if all its direct answers are necessarily true, and *foolish* if none of them is even possibly true. A special case is the *dumb* question, which has no direct answers. Here are examples:

1. Did you wear the black hat yesterday or did you wear the white one?
2. Did you wear a hat which is both black and not black, or did you wear one which is both white and not white?
3. What are three distinct examples of primes among the following numbers: 3, 5?

Clearly 3 is dumb and 2 is foolish. If we correspondingly call a necessarily false statement foolish too, we obtain the theorem *Ask a foolish question and get a foolish answer.* This was first proved by Belnap, but attributed by him to an early Indian philosopher mentioned in Plutarch's *Lives* who had the additional distinction of being an early nudist. Note that a foolish question contains all questions, and an empty one is contained in all.

Example 1 is there partly to introduce the question form used in 2, but also partly to introduce the most important semantic concept after that of direct answer, namely presupposition. It is easy to see that the two direct answers to 1 ('I wore the black hat', 'I wore the white one') could both be false. If that were so, the respondent would presumably say 'Neither', which is an answer not yet captured by our typology. Following Belnap who clarified this subject completely, let us introduce the relevant concepts as follows:

> a *presupposition*[40] of question Q is any proposition which is implied by all direct answers to Q.
> a *correction* (or *corrective answer*) to Q is any denial of any presupposition of Q.
> the (*basic*) *presupposition* of Q is the proposition which is true if and only if some direct answer to Q is true.

In this last notion, I presuppose the simplifying hypothesis which identifies a proposition through the set of worlds in which it is true; if that hypothesis is rejected, a more complex definition needs to be given. For example 1, 'the' presupposition is clearly the proposition that the addressee wore either the black hat or the white one. Indeed, in any case in which the number of direct answers is finite, 'the' presupposition is the disjunction of those answers.

Let us return momentarily to the typology of answers. One important family is that of the partial answers (which includes direct and complete answers). A second important family is that of the corrective answer. But there are still more. Suppose the addressee of question 1 answers 'I did not wear the white one.' This is not even a partial answer, by the definition given: neither direct answer implies it, since she might have worn both hats yesterday, one in the afternoon and one in the evening, say. However, since the questioner is presupposing that she wore at least one of the two, the response is *to him* a complete answer. For the response plus the presupposition together entail the direct answer that she wore the black hat. Let us therefore add:

> a *relatively complete answer* to Q is any proposition which, together with the presupposition of Q, implies some direct answer to Q.

We can generalize this still further: a complete answer to Q, relative to theory T, is something which together with T, implies some direct answer to Q—and so forth. The important point is, I think, that we should regard the introduced typology of answers as open-ended, to be extended as needs be when specific sorts of questions are studied.

Finally, which question is expressed by a given interrogative? This is highly context-dependent, in part because all the usual indexical terms appear in interrogatives. If I say 'Which one do you want?' the context determines a range of objects over which my 'which one' ranges—for example, the set of apples in the basket on my arm. If we adopt the simplifying hypothesis discussed above, then the main task of the context is to delineate the set of direct answers. In the 'elementary questions' of Belnap's theory ('whether-questions' and 'which-questions') this set of direct answers is specified through two factors: a *set of alternatives* (called the *subject* of the question) and *request* for a selection among these alternatives and, possibly, for certain information about the selection made ('distinctness and completeness claims'). What those two factors are may not be made explicit in the words used to frame the inter-rogative, but the context has to determine them exactly if it is to yield an interpretation of those words as expressing a unique question.

§4.3 *A Theory of Why-questions*

There are several respects in which why-questions introduce genuinely new elements into the theory of questions.[41] Let us focus first on the determination of exactly what question is asked, that is, the contextual specification of factors needed to understand a why-interrogative. After that is done (a task which ends with the delineation of the set of direct answers) and as an independent enterprise, we must turn to the evaluation of those answers as good or better. This evaluation proceeds with reference to the part of science accepted as 'background theory' in that context.

As example, consider the question 'Why is this conductor warped?' The questioner implies that the conductor is warped, and is asking for a reason. Let us call the proposition that the conductor is warped the *topic* of the question (following Henry Leonard's terminology, 'topic of concern'). Next, this question has a *contrast-class*, as we saw, that is, a set of alternatives. I shall take this contrast-class, call it X, to be a class of propositions which includes the topic. For this particular interrogative, the contrast could be that it is *this* conductor rather than *that* one, or that this conductor has warped rather than retained its shape. If the question is 'Why does this material burn yellow' the contrast-class could be the set of propositions: this material burned (with a flame of) colour x.

Finally, there is the respect-in-which a reason is requested, which determines what shall count as a possible explanatory factor, the relation of *explanatory relevance*. In the first example, the request might be *for events 'leading up to' the warping*. That allows as relevant an account of human error, of switches being closed or moisture condensing in those switches, even spells cast by witches (since the evaluation of what is a good answer comes later). On the other hand, the events leading up to the warping might be well known, in which case the request is likely to be for the standing conditions that made it possible for those events to lead to this warping: the presence of a magnetic field of a certain strength, say. Finally, it might already be known, or considered immaterial, exactly how the warping is produced, and the question (possibly based on a misunderstanding) may be about exactly what function this warping fulfils in the operation of the power station. Compare 'Why does the blood circulate through the body?' answered (1) 'because the heart pumps the blood through the arteries' and (2) 'to bring oxygen to every part of the body tissue'.

In a given context, several questions agreeing in topic but differing in contrast-class, or conversely, may conceivably differ further in what counts as explanatorily relevant. Hence we cannot properly ask what is relevant to this topic, or what is relevant to this contrast-class. Instead we must say of a given proposition that it is or is not relevant (in this context) to the topic with respect to that contrast-class. For example, in the same context one might be curious about the circumstances that led Adam to eat the apple rather than the pear (Eve offered him an apple) and also about the motives that led him to eat it rather than refuse it. What is 'kept constant' or 'taken as given' (that he ate the fruit; that what he did, he did to the apple) which is to say, the contrast-class, is not to be dissociated entirely from the respect-in-which we want a reason.

Summing up then, the why-question Q expressed by an interrogative in a given context will be determined by three factors:

The *topic* P_k
The *contrast-class* $X = \{P_1, \ldots, P_k, \ldots\}$
The *relevance relation R*

and, in a preliminary way, we may identify the abstract why-question with the triple consisting of these three:

$$Q = \langle P_k, X, R \rangle$$

A proposition A is called *relevant to* Q exactly if A bears relation R to the couple $\langle P_k, X \rangle$.

We must now define what are the direct answers to this question. As a beginning let us inspect the form of words that will express such an answer:

(*) P_k *in contrast to* (the rest of) X *because A*

This sentence must express a proposition. What proposition it expresses, however, depends on the same context that selected Q as the proposition expressed by the corresponding interrogative ('Why P_k?'). So some of the same contextual factors, and specifically R, may appear in the determination of the proposition expressed by (*).

What is claimed in answer (*)? First of all, that P_k is true. Secondly, (*) claims that the other members of the contrast-class are not true. So much is surely conveyed already by the question—it does not make sense to ask why Peter rather than Paul has paresis if they both have it. Thirdly, (*) says that A is true. And finally, there is that word 'because': (*) claims that A is a *reason*.

This fourth point we have awaited with bated breath. Is this not where the inextricably modal or counterfactual element comes in? But not at all; in my opinion, the word 'because' here signifies only that A is relevant, in this context, to this question. Hence the claim is merely that A bears relation R to $\langle P_k, X \rangle$. For example, suppose you ask why I got up at seven o'clock this morning, and I say 'because I was woken up by the clatter the milkman made'. In that case I have interpreted your question as asking for a sort of reason that at least includes events-leading-up-to my getting out of bed, and my word 'because' indicates that the milkman's clatter was that sort of reason, that is, one of the events in what Salmon would call the causal process. Contrast this with the case in which I construe your request as being specifically for a motive. In that case I would have answered 'No reason, really. I could easily have stayed in bed, for I don't particularly want to do anything today. But the milkman's clatter had woken me up, and I just got up from force of habit I suppose.' In this case, I do not say 'because' for the milkman's clatter does not belong to the relevant range of events, as I understand your question.

It may be objected that 'because A' does not only indicate that A is *a* reason, but that it is *the* reason, or at least that it is a good reason. I think that this point can be accommodated in two ways. The first is that the relevance relation, which specifies what sort of thing is being requested as answer, may be construed quite strongly: 'give me a motive strong enough to account for murder', 'give me a statistically relevant preceding event not screened off by other events', 'give me a common cause', etc. In that case the claim that the proposition expressed by A falls in the relevant range, is already a claim that it provides a telling reason. But more likely, I think, the request need not be construed that strongly; the point is rather that anyone who answers a question is in some sense tacitly claiming to be giving a good answer. In either case, the determination of whether the answer is indeed good, or telling, or better than other answers that might have been given, must still be carried out, and I shall discuss that under the heading of 'evaluation'.

As a matter of regimentation I propose that we count (*) as a direct answer *only if A is relevant*.[42] In that case, we don't have to understand the claim that A is relevant as explicit part of the answer either, but may regard the word 'because' solely as a linguistic signal that the words uttered are intended to provide an answer to the why-question just asked. (There is, as always, the tacit claim of the respondent that what he is giving is a good, and hence a relevant answer—we just do not need to make this claim part of the answer.) The definition is then:

> B is a *direct answer* to question $Q = \langle P_k, X, R \rangle$ exactly if there is some proposition A such that A bears relation R to $\langle P_k, X \rangle$ and B is the proposition which is true exactly if (P_k; *and for all $i \neq k$, not P_i*; and A) is true

where, as before, $X = \{P_1, \ldots, P_k, \ldots\}$. Given this proposed definition of the direct answer, what does a why-question presuppose? Using Belnap's general definition we deduce:

> a why-question *presupposes* exactly that
> (a) its topic is true

(b) in its contrast-class, only its topic is true
(c) at least one of the propositions that bears its relevance rela-
    tion to its topic and contrast-class, is also true.

However, as we shall see, if all three of these presuppositions are
true, the question may still not have a *telling* answer.

Before turning to the evaluation of answers, however, we must
consider one related topic: when does a why-question arise? In the
general theory of questions, the following were equated: question Q
arises, all the presuppositions of Q are true. The former means that
Q is not to be rejected as mistaken, the latter that Q has some true
answer.

In the case of why-questions, we evaluate answers in the light of
accepted background theory (as well as background information)
and it seems to me that this drives a wedge between the two concepts. Of course, sometimes we reject a why-question because we
think that it has no true answer. But as long as we do not think
that, the question does arise, and is not mistaken, regardless of
what is true.

To make this precise, and to simplify further discussion, let us
introduce two more special terms. In the above definition of 'direct
answer', let us call proposition A the *core* of answer B (since the
answer can be abbreviated to '*Because A*'), and let us call the proposition that (P_k *and for all* $i \neq k$, *not* P_i) the *central presupposition*
of question Q. Finally, if proposition A is relevant to $\langle P_k, X \rangle$ let us
also call it relevant to Q.

In the context in which the question is posed, there is a certain
body K of accepted background theory and factual information.
This is a factor in the context, since it depends on who the
questioner and audience are. It is this background which determines whether or not the question arises; hence a question may
arise (or conversely, be rightly rejected) in one context and not in
another.

To begin, whether or not the question genuinely *arises*, depends
on whether or not K implies the central presupposition. As long
as the central presupposition is not part of what is assumed
or agreed to in this context, the why-question does not arise at
all.

Secondly, Q presupposes *in addition* that one of the propositions
A, relevant to its topic and contrast-class, is true. Perhaps K does

not imply that. In this case, the question will still arise, provided *K* does not imply that all those propositions are false.

So I propose that we use the phrase 'the question arises in this context' to mean exactly this: *K* implies the central presupposition, and *K* does not imply the denial of any presupposition. Notice that this is very different from 'all the presuppositions are true', and we may emphasize this difference by saying 'arises in context'. The reason we must draw this distinction is that *K* may not tell us which of the possible answers is true, but this *lacuna* in *K* clearly does not eliminate the question.

§4.4 *Evaluation of Answers*

The main problems of the philosophical theory of explanation are to account for legitimate rejections of explanation requests, and for the asymmetries of explanation. These problems are successfully solved, in my opinion, by the theory of why-questions as developed so far.

But that theory is not yet complete, since it does not tell us how answers are evaluated as telling, good, or better. I shall try to give an account of this too, and show along the way how much of the work by previous writers on explanation is best regarded as addressed to this very point. But I must emphasize, first, that this section is not meant to help in the solution of the traditional problems of explanation; and second, that I believe the theory of why-questions to be basically correct as developed so far, and have rather less confidence in what follows.

Let us suppose that we are in a context with background *K* of accepted theory plus information, and the question *Q* arises here. Let *Q* have topic *B*, and contrast-class $X = \{B, C, \ldots, N\}$. How good is the answer *Because A*?

There are at least three ways in which this answer is evaluated. The first concerns the evaluation of *A* itself, as acceptable or as likely to be true. The second concerns the extent to which *A favours* the topic *B* as against the other members of the contrast-class. (This is where Hempel's criterion of giving reasons to expect, and Salmon's criterion of statistical relevance may find application.) The third concerns the comparison of *Because A* with other possible answers to the same question; and this has three aspects. The first is whether *A* is more probable (in view of *K*); the second whether it favours the topic to a greater extent; and the third, whether it is made wholly or partially irrelevant by other answers that could be given. (To this third aspect, Salmon's considerations about *screening off* apply.) Each of these three main ways of evaluation needs to be made more precise.

The first is of course the simplest: we rule out *Because A* altogether if K implies the denial of A; and otherwise ask what probability K bestows on A. Later we compare this with the probability which K bestows on the cores of other possible answers. We turn then to favouring.

If the question why B rather than C, ..., N arises here, K must imply B and imply the falsity of C, ..., N. However, it is exactly the information that the topic is true, and the alternatives to it not true, which is irrelevant to how favourable the answer is to the topic. The evaluation uses only that part of the background information which constitutes the general theory about these phenomena, plus other 'auxiliary' facts which are known but which do not imply the fact to be explained. This point is germane to all the accounts of explanation we have seen, even if it is not always emphasized. For example, in Salmon's first account, A explains B only if the probability of B given A does not equal the probability of B *simpliciter*. However, if I know that A and that B (as is often the case when I say that B because A), then my *personal probability* (that is, the probability given all the information I have) of A equals that of B and that of B given A, namely 1. Hence the probability to be used in evaluating answers is not at all the probability given all my background information, but rather, the probability given some of the general theories I accept plus some selection from my data.[43] So the evaluation of the answer *Because A* to question Q proceeds with reference only to a certain part $K(Q)$ of K. How that part is selected is equally important to all the theories of explanation I have discussed. Neither the other authors nor I can say much about it. Therefore the selection of the part $K(Q)$ of K that is to be used in the further evaluation of A, must be a further contextual factor.[44]

If $K(Q)$ plus A implies B, and implies the falsity of C, ..., N, then A receives in this context the highest marks for favouring the topic B.

Supposing that A is not thus, we must award marks on the basis of how well A redistributes the probabilities on the contrast-class so as to favour B against its alternatives. Let us call the probability in the light of $K(Q)$ alone the *prior* probability (in this context) and the probability given $K(Q)$ plus A the *posterior* probability. Then A will do best here if the posterior probability of B equals 1. If A is not thus, it may still do well provided it shifts the mass of the probability function toward B; for example, if it raises the probability of B while lowering that of C, ..., N; or if it does not lower the probability of B while lowering that of some of its closest competitors.

I will not propose a precise function to measure the extent to which the posterior probability distribution favours B against its alternatives, as compared to the prior. Two factors matter: the minimum odds of B against C, ..., N, *and* the number of alternatives in C, ..., N to which B bears these minimum odds. The first should increase, the second decrease. Such an increased favouring of the topic against its alternatives is quite compatible with a decrease in the probability of the topic. Imagining a curve which depicts the probability distribution, you can easily see how it could be changed quite dramatically so as to single out the topic—as the tree that stands out from the wood, so to say—even though the new advantage is only a relative one. Here is a schematic example:

Why E_1 rather than $E_2, \ldots, E_{1000}$?
Because A.
$Prob\ (E_1) = \ldots = Prob\ (E_{10}) = 99/1000 = 0.099$
$Prob\ (E_{11}) = \ldots = Prob\ (E_{1000}) = 1/99,000 \doteq 0.00001$
$Prob\ (E_1/A) = 90/1000 = 0.090$
$Prob\ (E_2/A) = \ldots = Prob\ (E_{1000}/A) = 910/999,000 \simeq 0.001$

Before the answer, E_1 was a good candidate, but in no way distinguished from nine others; afterwards, it is head and shoulders above all its alternatives, but has itself lower probability than it had before.

I think this will remove some of the puzzlement felt in connection with Salmon's examples of explanations that lower the probability of what is explained. In Nancy Cartwright's example of the poison ivy ('Why is this plant alive?') the answer ('It was sprayed with defoliant') was statistically relevant, but did not redistribute the probabilities so as to favour the topic. The mere fact that the probability was lowered is, however, not enough to disqualify the answer as a telling one.

There is a further way in which A can provide information which favours the topic. This has to do with what is called Simpson's

Paradox; it is again Nancy Cartwright who has emphasized the importance of this for the theory of explanation (see n. 13 above). Here is an example she made up to illustrate it. Let H be 'Tom has heart disease'; S be 'Tom smokes'; and E, 'Tom does exercise'. Let us suppose the probabilities to be as follows:

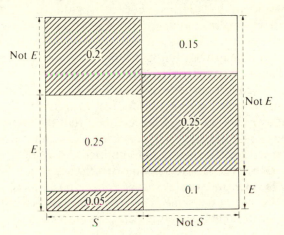

Shaded areas represent the cases in which H is true, and numbers the probabilities. By the standard calculation, the conditional probabilities are

Prob $(H/S) =$ *Prob* $(H) = \frac{1}{2}$
Prob $(H/S\&E) = \frac{1}{6}$
Prob $(H/E) = \frac{1}{8}$
Prob $(H/S$ & not $E) = 1$
Prob $(H/$ not $E) = \frac{3}{4}$

In this example, the answer 'Because Tom smokes' does favour the topic that Tom has heart disease, in a straightforward (though derivative) sense. For as we would say it, the odds of heart disease increase with smoking regardless of whether he is an exerciser or a non-exerciser, and he must be one or the other.

Thus we should add to the account of what it is for A to favour B as against $C, \ldots, N$ that: if $Z = \{Z_1, \ldots, Z_n\}$ is a logical partition of explanatorily relevant alternatives, and A favours B as against $C, \ldots, N$ if any member of Z is added to our background information. then A does favour B as against $C, \ldots, N$.

We have now considered two sorts of evaluation: how probable is *A* itself? *and*, how much does *A* favour *B* as against *C*, ..., *N*? These are independent questions. In the second case, we know what aspects to consider, but do not have a precise formula that 'adds them all up'. Neither do we have a precise formula to weigh the importance of how likely the answer is to be true, against how favourable the information is which it provides. But I doubt the value of attempting to combine all these aspects into a single-valued measurement.

In any case, we are not finished. For there are relations among answers that go beyond the comparison of how well they do with respect to the criteria considered so far. A famous case, again related to Simpson's Paradox, goes as follows (also discussed in Cartwright's paper): at a certain university it was found that the admission rate for women was lower than that for men. Thus 'Janet is a woman' appears to tell for 'Janet was not admitted' as against 'Janet was admitted'. However, this was not a case of sexual bias. The admission rates for men and women for each department in the university were approximately the same. The appearance of bias was created because women tended to apply to departments with lower admission rates. Suppose Janet applied for admission in history; the statement 'Janet applied in history' *screens off* the statement 'Janet is a woman' from the topic 'Janet was not admitted' (in the Reichenbach–Salmon sense of 'screens off': *P* screens off *A* from *B* exactly if the probability of *B* given *P* and *A* is just the probability of *B* given *P* alone). It is clear then that the information that Janet applied in history (or whatever other department) is a much more telling answer than the earlier reply, in that it makes that reply irrelevant.

We must be careful in the application of this criterion. First, it is not important if some proposition *P* screens off *A* from *B* if *P* is not the core of an answer to the question. Thus if the why-question is a request for information about the mechanical processes leading up to the event, the answer is no worse if it is statistically screened off by other sorts of information. Consider 'Why is Peter dead?' answered by 'He received a heavy blow on the head' while we know already that Paul has just murdered Peter in some way. Secondly, a screened-off answer may be good but partial rather than irrelevant. (In the same example, we know that there must be some true proposition of the form 'Peter received a blow on the head with impact *x*', but that does not disqualify the answer, it only means that some more informative answer is possible.) Finally, in the case of a deterministic process in which state A_i, and no other state, is followed by state A_{i+1}, the best answers to the question 'Why is the system in state A_n at time t_n?' may all have the form 'Because the system was in state A_i at time t_i', but each such answer is screened off from the event described in the topic by some other, equally good answer. The most accurate conclusion is probably no more than that if one answer is screened off by another, and not conversely, then the latter is better in some respect.

When it comes to the evaluation of answers to why-questions, therefore, the account I am able to offer is neither as complete nor as precise as one might wish. Its shortcomings, however, are shared with the other philosophical theories of explanation I know (for I have drawn shamelessly on those other theories to marshal these criteria for answers). And the traditional main problems of the theory of explanation are solved not by seeing what these criteria are, but by the general theory that explanations are answers to why-questions, which are themselves contextually determined in certain ways.

§4.5 Presupposition and Relevance Elaborated

Consider the question 'Why does the hydrogen atom emit photons with frequencies in the general Balmer series (only)?' This question presupposes that the hydrogen atom emits photons with these frequencies. So how can I even ask that question unless I believe that theoretical presupposition to be true? Will my account of why-questions not automatically make scientific realists of us all?

But recall that we must distinguish carefully what a theory *says* from what we believe when we accept that theory (or rely on it to predict the weather or build a bridge, for that matter). The epistemic commitment involved in accepting a scientific theory, I have argued, is not belief that it is true but only the weaker belief that it is empirically adequate. In just the same way we must distinguish what the question says (i.e. *presupposes*) from what we believe when we ask that question. The example I gave above is a question which arises (as I have defined that term) in any context in which those hypotheses about hydrogen, and the atomic theory in question, are *accepted*. Now, when I ask the question, if I ask it seriously and in my own person, I imply that I believe that this question arises. But that means then only that my epistemic commitment indicated by, or involved in, the asking of this question, is exactly—no more and no less than—the epistemic commitment involved in my acceptance of these theories.

Of course, the discussants in this context, in which those theories are accepted, are conceptually immersed in the theoretical world-picture. They talk the language of the theory. The phenomenological distinction between objective or real, and not objective or unreal, is a distinction between what there is and what there is not which is drawn in that theoretical picture. Hence the questions asked are asked in the theoretical language—how could it be otherwise? But the epistemic commitment of the discussants is not to be read off from their language.

Relevance, perhaps the other main peculiarity of the why-question, raises another ticklish point, but for logical theory. Suppose, for instance, that I ask a question about a sodium sample, and my background theory includes present atomic physics. In that case the answer to the question may well be something like: because this material has such-and-such an atomic structure. Recalling this answer from one of the main examples I have used to illustrate the asymmetries of explanation, it will be noted that, *relative to* this background theory, my answer is a proposition necessarily equivalent to: because this material has such-and-such a characteristic spectrum. The reason is that the spectrum is unique —it identifies the material as having that atomic structure. But, here is the asymmetry, I could not well have answered the question by saying that this material has that characteristic spectrum.

These two propositions, one of them relevant and the other not, are equivalent relative to the theory. Hence they are true in exactly the same possible worlds allowed by the theory (less metaphysically put: true in exactly the same models of that theory). So now we have come to a place where there is a conflict with the simplifying hypothesis generally used in formal semantics, to the effect that propositions which are true in exactly the same possible worlds are identical. If one proposition is relevant and the other not, they cannot be identical.

We could avoid the conflict by saying that of course there are possible worlds which are not allowed by the background theory. This means that when we single out one proposition as relevant, in this context, and the other as not relevant and hence distinct from the first, we do so in part by thinking in terms of worlds (or models) regarded in this context as impossible.

I have no completely telling objection to this, but I am inclined to turn, in our semantics, to a different modelling of the language, and reject the simplifying hypothesis. Happily there are several sorts of models of language, not surprisingly ones that were constructed in response to other reflections on relevance, in which propositions can be individuated more finely. One particular sort of model, which provides a semantics for Anderson and Belnap's logic of tautological entailment, uses the notion of *fact*.[45] There one can say that

> It is either raining or not raining
> It is either snowing or not snowing

although true in exactly the same possible situations (namely, in all) are yet distinguishable through the consideration that today, for example, the first is *made true* by the fact that it is raining, and the second is made true by quite a different fact, namely, that it is not snowing. In another sort of modelling, developed by Alasdair Urquhart, this individuating function is played not by facts but by bodies of information.[46] And still further approaches, not necessarily tied to logics of the Anderson–Belnap stripe, are available.

In each case, the relevance relation among propositions will derive from a deeper relevance relation. If we use facts, for example, the relation R will derive from a request to the effect that the answer must provide a proposition which describes (is made true by) facts of a certain sort: for example, facts about atomic structure, or facts about this person's medical and physical history, or whatever.

§5. *Conclusion*

Let us take stock. Traditionally, theories are said to bear two sorts of relation to the observable phenomena: *description* and *explanation*. Description can be more or less accurate, more or less informative; as a minimum, the facts must 'be allowed by' the theory (fit some of its models), as a maximum the theory actually implies the facts in question. But in addition to a (more or less informative) description, the theory may provide an explanation. This is something 'over and above' description; for example, Boyle's law describes the relationship between the pressure, temperature, and volume of a contained gas, but does not explain it—kinetic theory explains it. The conclusion was drawn, correctly I think, that even if two theories are strictly empirically equivalent they may differ in that one can be used to answer a given request for explanation while the other cannot.

Many attempts were made to account for such 'explanatory power' purely in terms of those features and resources of a theory that make it informative (that is, allow it to give better descriptions). On Hempel's view. Boyle's law does explain these empirical facts about gases, but minimally. The kinetic theory is perhaps better *qua* explanation simply because it gives so much more information about the behaviour of gases, relates the three quantities in question to other observable quantities, has a beautiful simplicity, unifies our over-all picture of the world, and so on. The use of more sophisticated statistical relationships by Wesley Salmon and James Greeno (as well as by I. J. Good, whose theory of such concepts as weight of evidence, corroboration, explanatory power, and so on deserves more attention from philosophers), are all efforts along this line.[47] If they had succeeded, an empiricist could rest easy with the subject of explanation.

But these attempts ran into seemingly insuperable difficulties. The conviction grew that explanatory power is something quite irreducible, a special feature differing in kind from empirical adequacy and strength. An inspection of examples defeats any attempt to identify the ability to explain with any complex of those more familiar and down-to-earth virtues that are used to evaluate the theory *qua* description. Simultaneously it was argued that what science is really after is understanding, that this consists in being in a position to explain, hence what science is really after goes well beyond empirical adequacy and strength. Finally, since the theory's ability to explain provides a clear reason for accepting it, it was argued that explanatory power is evidence for the *truth* of the theory, special evidence that goes beyond any evidence we may have for the theory's empirical adequacy.

Around the turn of the century, Pierre Duhem had already tried to debunk this view of science by arguing that explanation is not an aim of science. In retrospect, he fostered that explanation–mysticism which he attacked. For he was at pains to grant that explanatory power does not consist in resources for description. He argued that only metaphysical theories explain, and that metaphysics is an enterprise foreign to science. But fifty years later, Quine having argued that there is no demarcation between science and philosophy, and the difficulties of the ametaphysical stance of the positivist-oriented philosophies having made a return to metaphysics tempting, one noticed that scientific activity does involve explanation, and Duhem's argument was deftly reversed.

Once you decide that explanation is something irreducible and special, the door is opened to elaboration by means of further concepts pertaining thereto, all equally irreducible and special. The premisses of an explanation have to include lawlike statements; a statement is lawlike exactly if it implies some non-trivial counter-factual conditional statement; but it can do so only by asserting relationships of necessity in nature. Not all classes correspond to genuine properties; properties and propensities figure in explanation. Not everyone has joined this return to essentialism or neo-Aristotelian realism, but some eminent realists have publicly explored or advocated it.

Even more moderate elaborations of the concept of explanation make mysterious distinctions. Not every explanation is a scientific explanation. Well then, that irreducible explanation-relationship comes in several distinct types, one of them being scientific. A scientific explanation has a special form, and adduces only special sorts of information to explain—information about causal connections and causal processes. Of course, a causal relationship is just what 'because' must denote; and since the *summum bonum* of science is explanation, science must be attempting even to describe something beyond the observable phenomena, namely causal relationships and processes.

These last two paragraphs describe the flights of fancy that become appropriate if explanation is a relationship *sui generis* between theory and fact. But there is no direct evidence for them at all, because if you ask a scientist to explain something to you, the information he gives you is not different in kind (and does not sound or look different) from the information he gives you when you ask for a description. Similarly in 'ordinary' explanations: the information I adduce to explain the rise in oil prices, is information I would have given you to a battery of requests for description of oil supplies, oil producers, and oil consumption. To call an explanation scientific, is to say nothing about its form or the sort of information adduced, but only that the explanation draws on science to get this information (at least to some extent) and, more importantly, that the criteria of evaluation of how good an explanation it is, are being applied using a scientific theory (in the manner I have tried to describe in Section 4 above).

The discussion of explanation went wrong at the very beginning when explanation was conceived of as a relationship like description: a relation between theory and fact. Really it is a three-term relation, between theory, fact, and context. No wonder that no single relation between theory and fact ever managed to fit more than a few examples! Being an explanation is essentially relative, for an explanation is an *answer*. (In just that sense, being a daughter is something relative: every woman is a daughter, and every daughter is a woman, yet being a daughter is not the same as being a woman.) Since an explanation is an answer, it is evaluated *vis-à-vis* a question, which is a request for information. But exactly what is requested, by means of the interrogative 'Why is it the case that *P*?', differs from context to context. In addition, the background theory plus data relative to which the question is evaluated, as arising or not arising, depends on the context. And even what part of that background information is to be used to evaluate how good the answer is, *qua* answer to that question, is a contextually determined factor. So to say that a given theory can be used to explain a certain fact, is always elliptic for: there is a proposition which is a telling answer, relative to this theory, to the request for information about certain facts (those counted as relevant for *this* question) that bears on a comparison between this fact which is the case, and certain (contextually specified) alternatives which are not the case.

So scientific explanation is not (pure) science but an application of science. It is a use of science to satisfy certain of our desires; and these desires are quite specific in a specific context, but they are always desires for descriptive information. (Recall: every daughter is a woman.) The exact content of the desire, and the evaluation of how well it is satisfied, varies from context to context. It is not a single desire, the same in all cases, for a very special sort of thing, but rather, in each case, a different desire for something of a quite familiar sort.

Hence there can be no question at all of explanatory power as such (just as it would be silly to speak of the 'control power' of a theory, although of course we rely on theories to gain control over nature and circumstances). Nor can there be any question of explanatory success as providing evidence for the truth of a theory that goes beyond any evidence we have for its providing an adequate description of the phenomena. For in each case, a success of explanation is a success of adequate and informative description. And while it is true that we seek for explanation, the value of this search for science is that the search for explanation is *ipso facto* a search for empirically adequate, empirically strong theories.

6

Probability: the New Modality of Science

> The majority of men follow their passions, which are
> movements of the sensitive appetite, in which movements
> heavenly bodies can cooperate; but few are wise enough
> to resist these passions. Consequently astrologers are able
> to foretell the truth in the majority of cases, especially in a
> general way. But not in particular cases ...
>
> St. Thomas Aquinas, *Summa Theologiae*
> 1, Qu. 115, a.4, *ad* Obj. 3

IN the Aristotelian tradition, natural philosophy was deeply involved
with modality: necessity, possibility, contingency, potentiality.
Nominalism and modern empiricism rejected this role of modality,
arguing that necessity, for example, attaches only to relations among
ideas, or among words, and not to physical occurrences. But
empiricism has not been lucky with modality; as Herman Weyl
said, the ghost of modality is not easily laid.[1] In this century, the
problem has become much more acute, for a new modality, a
possibility-with-degrees, has taken centre stage in physical science:
probability.

Scientific realists are dealing with modalities by reifying certain
corresponding 'entities'. Thus in the philosophy of space and time,
where possible light ray paths and possible trajectories of moving
bodies play an important role, it has been asserted that space-time
is itself a real, substantial, concrete entity.[2] In that case, a possible
light ray path is a real part of that real entity—a geodesic curve—
and the talk of possibility has been effectively eliminated. In philosophy of quantum mechanics there has been the suggestion by
Everett that all the worlds in the Everett–de Witt 'many worlds
interpretation' be regarded as real.[3] In general philosophy of
science, David Lewis has advanced a view of laws of nature as
factual statements about the real possible worlds (as opposed to the
logically conceivable but unreal worlds) of which ours, the actual

one, is but an equal among equals.[4] And finally, for probability, there is the propensity interpretation, according to which probability is itself a physical magnitude, the strength or intensity of the real chance of occurrence of an event, which cannot be eliminated through reference to actual classes of actual occurrences.[5]

Is a philosophical retrenchment possible for modality? In this chapter I shall concentrate on the special topic of probability, and only return briefly to the general question at the end. I shall argue that a constructive account of probability in physics is possible within the nominalist, empiricist tradition. This will be a long chapter, so here is a preview. Sections 1–4 are devoted to probability; and of these the first three, which concern the role of probabilities as they occur in scientific theories, are meant to show that both infinity and possibility need to be taken very seriously in any account of that role. This shows already, in a general way, that no simple empiricist account of probability is feasible. In Section 4 I shall show that if we accept probability theory in the form in which it is used in today's science, the probability of events cannot be identified with the relative frequency of the occurrence of actual events of the same sort. In that section I shall propose an amended frequency interpretation, in which probabilities, though not identified with frequencies, are still construed in terms of frequencies. But this is a modal account in that the only reasonable, non-technical statement of it is in terms of what would be, or could be, the case and not only of what is actually the case. Section 5 will then return to the question how an empiricist can view this apparently inescapable modal element in physical theory.

§1. *Statistics in General Science*

Probability is not found only in physics. The theory of probability is widely used in science today because it provides the foundation for statistics, and statistical methods have become a major tool in all the sciences, both pure and applied.

What exactly is the distinction between probability theory and statistics? Usage is not uniform; sometimes the two terms are used interchangeably. But it is quite clear, I think, what is meant by such a term as 'the working statistician'. Let me propose this line of division: *statistics* is the science that deals with distributions and proportions in actual (large but finite) classes (also called 'populations', 'aggregates', 'ensembles') of actual things. What is often called a *statistic* is a statement about such a distribution, such as

65% of all American males are grossly overweight.

Statistical methods are designed to arrive at such statistics on the basis of data concerning small samples, to test hypotheses implying such statistics, and to infer new statistics from given ones. In doing so, and in devising these methods, the statistician draws on the mathematical theory of probability.

That mathematical theory is not concerned solely with the large but finite classes which are the basic topic of concern for statistics, nor even limited to classes of things limited by what there actually is in the world. Extrapolation to infinities is the main device whereby probability theory has advanced the cause of statistics.

This role of probability does not create philosophical problems. Nor does the reporting of statistics in probability terminology, as when statistical findings are reported as 'American males are more likely to be overweight than Eskimos', or 'Jones is an American male. Statistics show that he is probably overweight, lives in the suburbs, drives to work, ...'

There is some appearance in statistics of dealing with infinite classes, but this appearance is deceptive. The reason is that statistical methods are more reliable the larger the populations to which they are applied. This can be illustrated with the quotation at the beginning of this chapter from Aquinas, who seems to have regarded astrology as a (successful) statistical science. Exactly what class of things was he talking about when he said that astrologers manage to foretell the truth 'in the majority of cases'? We can guess that he was referring at least to the class of astrological predictions made in his own century. But because he had an explanation of their success, he no doubt expected astrology to remain successful, so his claim can also be taken to extend to the class of predictions in the period 1000 to 1500, the period 900 to 1600, and so on. On the other hand he probably did not mean that on every given day, most of the astrological predictions made are true—temporary fluctuations in the success rate are possible, though less likely the larger the daily number of astrological predictions becomes.

So the 'majority of cases' can be construed as referring to 'the long run' of all astrological predictions. That class, however, may be infinite—namely, if the human race does not die out, and if past experience is any indication of the popularity of astrology. So is Aquinas's statistical claim an assertion about an infinite class after all? The answer is that we need not so understand it, for we can take it to be a complex general claim about a series of finite classes of increasing size. We can construe

> in the long run, the majority of astrological predictions will be true

as saying,

> the proportion T_t of truths among the astrological predictions made before time t, converges to a number greater than $\frac{1}{2}$

where the little letter t ranges over the positive numbers. (That number to which the series T_t converges would then be called the *relative frequency* of astrological success in the long run.)

To sum up then, statistics as such is concerned with statements of proportion or distribution, in actual, finite classes, and these raise no philosophical perplexities. If the uses of the concept of probability were restricted to statistical calculations, we could rest easy with it too—and most of its uses are. But not all.

§2. *Classical Statistical Mechanics*

Probability theory was mainly developed in the eighteenth century; its application to physics followed in the nineteenth. Statistical mechanics, at the hands of Maxwell, Boltzmann, and Gibbs, extended mechanics to the theory of heat and the general theory of gases—the phenomena of thermodynamics. When we inspect the use of probability here, we cannot easily assimilate it to statistics in the strict sense of the preceding section. In expositions of the new statistical approach in physics, two intuitive notions were used: probability as *degree of ignorance* and probability as *measure of objective quantities* such as frequencies of occurrence, averages, time of sojourn. We shall have to disentangle these two notions.

§2.1 *The Measure of Ignorance*

Poincaré's expositions of the use of probability in physics relied on examples about gambling devices, such as the roulette wheel.[6] It was in the study of gambling that probability theory had originated, and such examples are indeed apt for the exposition of nineteenth-century statistical mechanics. Suppose that the croupier gives an impulse to the gambling wheel, by means of his hand applied to point x on the wheel. What will be the final rest position of that point? If we knew the exact initial position, the exact impulse, the exact frictional forces, and so forth, the laws of classical mechanics would in principle suffice to deduce an exact final position. But we do not have such exact knowledge—we only know the values of the initial quantities approximately. Hence, if we are to have a mechanics of practical use, we must devise a method of calculating all the information derivable, via the laws of mechanics, from an approximate specification of the initial values.

The basic conceptual advance in this problem occurred when it was realized that 'approximate' is misleadingly vague, and can be replaced by a quantitative concept. Suppose I begin by saying: the initial impulse the croupier gave the wheel was approximately 1. Then I improve on that by saying: it was $1 \pm d$. This means: the exact impulse is a number in the interval $(1 - d, 1 + d)$. I could arrive at this judgement by asking the croupier to submit himself to a series of measurements with a suitable 'impulse meter' and noting that the impulses measured in this series all fell in that interval.

But if this is all I do, I am discarding information. For the measurements on the croupier reveal that those impulses are distributed in the interval $(1 - d, 1 + d)$ in an uneven fashion. Most lie close to 1; and the few that lie close to $1-d$ are closely matched in number by those close to $1 + d$. I should conclude that not every impulse in this interval is equally likely: instead, they fit into what is called a *normal distribution* around 1.

This term *normal distribution* is taken from probability theory, and the *quantitative measure of our ignorance* about the exact impulse on a given occasion, at which we have now arrived, is a subject of probability theory.

Using this quantitatively expressed combination of knowledge and ignorance about the initial values, plus the laws of mechanics that relate to exact initial values, we derive a similarly expressed conclusion about the final values. That is, we deduce that the point on the wheel will come to rest at a position in the interval $(q - k, q + k)$, but those positions are not all equally likely: their likelihood clusters around q, and indeed, the distribution is normal.

Let me add a second example. I know about Jean-Paul Jones only that he was drafted to the infantry in 1944, and I am interested in what he is (like) today. Well, there is a great deal of relevant information. The infantry files tell us that among those draftees, there was a certain distribution d_1 of ages between 18 and 24, a certain distribution d_2 of heights between 5 feet 7 inches and 6 feet 2 inches, a distribution d_3 of weights between 120 and 200 pounds. These distributions are not exactly normal distributions, but not far from it. Meanwhile, insurance companies have information about the 'dynamics' of these quantities—specifically about termination by death, and change of weight with age in this population. From all this I deduce that Mr Jones is 'most likely' alive today, with age close to 56, and height (still) near 5 feet 10 inches, weight increased to close to 180 pounds. This 'most likely' is the deceptive shorthand for the fact that we have a good deal of information about the distribution d'_i of those quantities 34 years later. Thus we have here too a quantitative measure of ignorance of initial conditions dynamically transformed to a corresponding measure of ignorance of final conditions.

After these two simple examples, it will be easy to explain Liouville's Theorem, a central result in statistical mechanics. Consider an isolated mechanical system with known total energy E, but unknown mechanical state S_t at time t. The possible states are represented by points in a space, called the phase-space, whose co-ordinates are the position and momentum coordinates of the constituent molecules of this system. Hence, designating these co-ordinates as $x_1, \ldots x_n$, we write

(1) $S_t = (x_1(t), \ldots x_n(t))$
(2) $E = H(x_1, \ldots x_n)$—constant

since the energy is a function of the state, and the state itself a function of the time.

The region of points in phase-space that satisfy equation (2) is called an *energy surface*; and as the state S_t changes with time we can picture to ourselves this system travelling around on that energy surface. Let us now introduce what we do and do not know about the state at t in the quantitative fashion learned from probability theory. Let $P_t(X)$ be the probability that the state S_t of the system at time t is in region X. What that probability is, simply summarizes our information; for example, we might know that the system is a gas kept inside a container, so that the position coordinates are limited by the positions of the walls of that container.

This probability function is formally rather like mass; we can express it in the form: density × volume. So there is a probability density function P_t. Liouville's Theorem now says that the probability transforms with time in a way that is characterized by a constant density $(dP_t/dt = 0)$. This is how the determinism of classical physics shows up in statistical thermodynamics: for each initial state x in region X, the laws of mechanics determine a unique final state x' after an interval of time of length m, say (so that, if $S_t = x$ then $S_{t+m} = x'$) and this 'flow' of the possible states through region X is formally like the flow of an incompressible fluid, in that its density does not change. So for example, if we have for our initial probability P_t a uniform distribution over a region X, then the final probability P_{t+m} is still a uniform distribution—though now over the 'image' of X, that is, the region of points x' related through the state transformation to points x in X.

§2.2 Objective and Epistemic Probability Disentangled[7]

Summarizing the discussion so far, we may say that classical statistical mechanics is just classical mechanics, applied under conditions of less than perfect information. This sort of application was raised from the level of plausible remarks about approximation to one of sophisticated quantitative analysis through the application of probability theory—the study of quantitative measures of ignorance developed in connection with gambling.

But still, statistical mechanics is a branch of physics. It was developed in its own right, often with only a precarious hold on its supposed deterministic underpinnings. Has physics then become in part a study of human ignorance, an amalgam of subjective and objective factors? This view of the matter fits uneasily with other reflections on gas theory. If the scientist switches in discussion from the kinetic energies of individual molecules, to the mean kinetic energy of an ensemble of molecules—has he begun to study a subject which involves essential reference to ignorance? Is the mean kinetic energy of the ensemble not as objective a fact in the world as the kinetic energy of any one individual molecule? Should we not be saying rather that the gambling history of probability theory brought an essentially irrelevant subjective terminology into physics?

These questions express the puzzlement, the tension, that surrounds the interpretation of probability in physics, and elsewhere, and which gives rise to competing philosophical views on what probability is. In the case which we are at present examining, we must carefully disentangle the objective and subjective factors. Let us look back for a moment to the general use of statistics, and our example of Jean-Paul Jones, the infantry draftee. The statistical information about the infantrymen—the *statistics*—is purely objective. But it is related to our subjective uncertainty about Jones. The paradigm relation between the two is located in what we call the *statistical syllogism*:

1. 73 per cent of the infantrymen of 1944 are still alive.
2. Jones was an infantryman in 1944.
3. I have no other information about Jones relevant to the question whether he is still alive.
4. Therefore, the probability (for me) that Jones is still alive, equals 0.73.

The *statistic* is the first premiss. Both premisses 1 and 2 are purely objective, and are not about belief, knowledge, ignorance, or uncertainty in any way at all. But the conclusion, 4, is about information and the lack thereof—namely, mine—and it manages to be because the third premiss said something about my state of information.

Perhaps the term 'probability' has more than one sense. The sense in which it occurs in the above conclusion we may designate as *epistemic probability*. This is not a sort of probability a proposition can have in and by itself, or in relation to the facts that it is about —the epistemic probability of a proposition is a relation it bears to a given person, or more strictly, to a body of information (that person's information). If I say 'The probability that Jones is still alive equals 0.73', and mean it in this sense, then I am summarizing, in a precise and effective manner, the net total of my information concerning whether or not Jones is still alive.

The information which I am summarizing, however, does not have the word 'probability' in it; or, at least, it need not. In this particular case the information is that a certain individual belongs to a certain class, and that the proportion of survivors in that class (in 1978) equals 73 per cent.

In just the same way, statistical mechanics has, *in itself*, nothing to do with human ignorance. It is related to epistemic probability only by supplying objective premises for such statistical syllogisms and similar inferences.

Consider the usual body of gas in a container. Its *macrostate* is specified by giving the values of such macroscopic quantities as volume, temperature, and pressure. These determine the total energy of the ensemble of molecules with which this gas is theoretically identified. The *microstate* is the mechanical state of this ensemble of molecules, comprising all the individual positions and momenta. To simplify the discussion temporarily, let us speak as if to each macrostate there corresponds a finite number of possible microstates. Maxwell and Boltzmann introduced the postulate that all these microstates are equally probable.

What shall we make of this postulate? Does it concern epistemic probability, and say that we have no information that favours one microstate over another? Well, yes, but only indirectly; via a statistical syllogism. The second and third premisses in that syllogism are:

(2') The gas is in macrostate D at time t.

(3') I have no other information about the gas relevant to the question of what microstate it is in at time t.

To get to the conclusion of epistemic equi-probability, we now need an objective first premiss. That is exactly the proposition postulated. While the gas remains in macrostate D, its microstate is continuously changing, since its molecules are in motion. Continuing the fiction that there are only finitely many microstates, that first premiss reads:

(1') While a gas is in macrostate D, it spends as much time in one microstate compatible with D as in any other (*Ergodic Hypothesis*).

The time spent in a state is called *sojourn time* and the objective information given is the equality of sojourn times.

We have now come to an objective quantity measured by a probability function—the proportion of time of sojourn. This is comparable to a statistic. We must keep firmly in mind that to call a measure of something a probability function, is not to say that this something is objective *or* subjective. For a probability function can be a measure of ignorance, or of proportions in a population, or of proportions of times of sojourn. What is important, is that in the science of statistical mechanics, one deals *directly* with measures of objective quantities. These are related, via statistical syllogisms and similar inferences, to judgements of epistemic probability. Because the relation between the two can be so intimate—note that the number that appears in premiss 1 and conclusion 4 is really the same number—it would generally be pedantic and a hindrance for the scientist to keep distinguishing them in thought and terminology. But if we are not to fall prey to confusion here, we must keep firmly in mind that the Maxwell–Boltzmann equi-probability hypothesis is a hypothesis about equality of objective quantities, which have *in themselves* nothing to do with human ignorance.

§2.3 *The Intrusion of Infinity*

To facilitate the discussion, I pretended that only finitely many microstates are compatible with a macrostate. But of course, the microstate space (phase-space) is continuous. The equi-probability postulate is really that the system spends equally much time in sub-regions of equal volume, of the region corresponding to the macro-state. These subregions can be taken ever smaller, so as to converge on the points which represent the microstates. But because of the continuity, attention to infinities can no longer be avoided.

This gave rise to the second great chapter in the history of probability theory. Its initial motivation was the motivation of statistics proper: gambling, insurance, theory of errors, the census, genetics, and natural selection. In physics, probabilities were represented by functions of volume, and hence in mathematics, in the newly developing branch of *measure theory*, volume and probability became but two examples of those functions, mapping sets into real numbers, that are called 'measures'. For probability, this development culminated in the new axiomatic theory of Kolmogoroff, today the recognized foundation. There are still influential mathematicians and philosophers who regard Kolmogoroff's theory as dealing with only a special case in probability theory; but they too must admit that this special case is the special case of all uses of probability in standard physics.

Let me briefly explain these axiomatic foundations. To use a concrete example, suppose that a flea is jumping around on a table top. This table top is a spatial region, part of a plane; and any part of it can be regarded as a set of points that the flea can land on. If B is such a part of the table top, the event we are interested in is the event *that the flea lands in B*. Calling the flea t (short for 'Thumper'), we may call that event Bt.

Suppose A and B are two parts of the table top. We call them *disjoint* if they do not overlap; and the region comprising all the points in A plus all the points in B we call their *union*, $A \cup B$. The corresponding event can be designated either as $(A \cup B)t$ or as $At \cup Bt$—it is clearly the event that the point the flea lands on is either in A or in B.

Infinity comes in this way. Suppose that A^1, A^2, A^3, ... are infinitely many parts of the table top. They too have a union; and we write that as $\cup A^i$. Finally, the table top as a whole is a region too; call it K.

The general axioms by which the whole theory of probability may be characterized are

1. $0 \leqslant P(Bt) \leqslant 1$; $P(Kt) = 1$
2. If A and B are disjoint, then
   $P(At \cup Bt) = P(At) + P(Bt)$

The second axiom is called the axiom of *finite additivity*. But suppose the probability of At is proportional to the area of A. In that case, the probability of $(\cup A^i t)$ is proportional to the area of $\cup A^i$, which is the *infinite sum* of the areas of the parts A^i. This will indicate why Kolmogoroff replaced 2 by

2*. If all the A^i are mutually disjoint, then

$$P(\cup A^i t) = \sum_{i=1}^{\infty} P(A^i t)$$

which is called the axiom of *countable additivity*.

We can no longer say that all of probability can be explained as a theory about proportions in finite classes. For the theory with axioms 1 and 2* has models which are radically different from (not isomorphic to) any model in which all the probabilities correspond to such finite proportions. At this point this reflection may seem to be merely of mathematical interest, but as we shall see, it has created serious difficulties for the philosophical interpretation of probability.

However, we should also notice that the subject still concerns proportions, if not finite proportions. The proportion of the area of region A to the total area of the table top K, determines (if only in part) the probability that the flea will land in A. So although the proportion is one between quantities defined on sets of points (infinite sets clearly) and infinity enters into the manipulation of these sets, we are still dealing with a relatively straightforward extrapolation of the finite proportions dealt with in ordinary statistics.

§3. *Probability in Quantum Mechanics*

Probability appeared in atomic physics very early, but at first there was no reason to think that it brought complexities different from those in classical physics. Radio-active decay, for example, introduces probabilities. The heaviest elements in the periodic table are naturally radio-active, that is, emit radiation. Simultaneously, they transmute into lighter elements; and this transmutation is occurring spontaneously all the time. For example, *radium* transmutes into *radon*. The rate at which this transmutation occurs can be measured, and the finding is that if we have a macroscopic quantity of radium, half of it will become radon in 1,600 years. This period is therefore called the *half-life* of radium. (Thus in 3,200 years, one quarter of original quantity would still be radium.)

What I have just stated is a macroscopic law governing the temporal development of radium. If we take the atomic hypothesis, then that law can only be approximately correct. For if we divide the radium into ever smaller quantities, then after finitely many steps we arrive at single atoms. But it does not make sense to say that in 1,600 years, half of one atom, or half of an ensemble of atoms with only one member, turns into radon. Indeed, it does not make sense to say that about an ensemble containing one million and one, or any odd number, of atoms. So the statement relating to atoms is: the probability that a given radium atom decays and transmutes into radon within 1,600 years, equals $\frac{1}{2}$.

But this does not look unfamiliar. No doubt the macroscopic law can be stated as follows: within 1,600 years, half of *all* radium atoms decay. I have no information about *this* radium atom relevant to the question when it will decay, except that it is *a* radium atom. Therefore, the probability (to me) that this atom will decay within 1,600 years equals $\frac{1}{2}$.

If not unfamiliar, it is still worrying. What if there were only a few radium atoms in the world, or an odd number? If we change the above to 'approximately half of all radium atoms', should we also say that the probability is only approximately half? And if so, does that mean that there is a real but unknown probability, which is a real number very close to $\frac{1}{2}$, but cannot be further determined by us? Atomic theory says: the probability *is* $\frac{1}{2}$. If it says that so unambiguously, that cannot be epistemic probability concluded by means of the above statistical syllogism then. There is certainly a relation between the probability postulated and the proportion of all radium that decays in any 1,600 year period—but the relation does not look any too straightforward. Specifically, we cannot say: the probability is merely a measure of that proportion, for the former is exactly $\frac{1}{2}$ and the latter only approximately $\frac{1}{2}$.

In this way the mere fact of discreteness, of the granular structure of matter, seems to introduce problems for the interpretation of probability. However, to begin with there was every hope that upon better understanding of the phenomena, these problems would disappear. For in the past phenomena had been studied which exhibited discontinuous changes; but in the models constructed, that appearance had itself been identified with continuous change only approximately discontinuous. So could it not be expected that the deep theory behind the apparently discontinuous transmutation of radium into radon would identify probabilities with exact proportions of times of sojourn in (as yet unconceived) underlying states?

This hope was discarded in the now well-known, but spectacularly revolutionary break with tradition of the quantum theory. Probabilities in quantum theory cannot be glossed with probabilities in an underlying theory of the classical sort.

Before I continue the exposition, however, I should emphasize that in saying even this much, I am already somewhat partisan with respect to a certain sort of interpretation of quantum physics. This is still an area of lively philosophical debate. To be as fair as possible, I shall concentrate on aspects on which most of the extant interpretations agree, and if not that, state what is generally called the orthodox position—but total philosophical neutrality will, I think, be impossible.[8]

§3.1 *The Disanalogies with the Classical Case*

In quantum mechanics, we find a distinction which is formally similar to the classical one between macrostates and microstates— namely between *mixtures* (or, *mixed states*) and *pure states*. There are two main problems about these that involve probabilities: how mixtures are related to pure states, and how pure states are related to each other. I shall discuss these in turn.

Mixed states are typically introduced in situations of uncertainty.

Suppose that helium atoms are escaping through a small crack in an oven in a laboratory. These escaping atoms have different energies. So the ensemble of atoms is mixed, a mixture of atoms in different states. For simplicity let us assume that we are looking at some situation like this, in which particles escape from some device, each in one of three states w_1, w_2, w_3. Suppose also that the proportions of these three are equal. We can now also say that the device prepares the particles in the mixed state

1. $w = \frac{1}{3}w_1 + \frac{1}{3}w_2 + \frac{1}{3}w_3$

(For the technically minded: w_1, w_2, w_3, and w all stand here for statistical operators, or density matrices; not vectors.)

Because of the way I have introduced mixed states here (and I did so in standard textbook fashion), their interpretation looks very straightforward. To say that a system is in mixed state w, means only that it is really in one of the three pure states, though we don't know which; and the numbers attached to the components measure our ignorance (epistemic probability). This is called the *Ignorance Interpretation* of mixtures. It is the interpretation that seems to be implicit in many discussions of mixed states; and it was advanced explicitly in 1948 by Hans Reichenbach.[9] It is similar to the principle that a gas is in macrostate D exactly if it is in one of the microstates compatible with D.

Unfortunately there is the problem of *degeneracy*. When we decompose a given mixed state w into its pure components, there is in general more than one way to do it. So 1 is compatible with:

2. $w = \frac{1}{3}w_1 + \frac{1}{3}w_2' + \frac{1}{3}w_3'$

where w_2' and w_3' are totally different states from w_2 and w_3.

If we tried to follow through with the Ignorance Interpretation now, we would have to conclude that the probability that a given one of these particles is in state w equals $1 + \frac{2}{3}$, since we have to add up the probabilities of incompatible (disjoint) events.

That this really does not make sense will perhaps be clearer if we look at what we may call the naïve statistical interpretation. This says that the state w cannot be ascribed to any individual particle, but only to the ensemble of particles; and that the number $\frac{1}{3}$ measures the relative size of the sub-ensemble which comprises those of the particles which are in pure state w_1, and similarly for w_2 and w_3. But then we must say the same for w_2' and w_3'. Since

$\frac{1}{3}$ taken five times is more than one, these five sub-ensembles must overlap. But they cannot! If they overlapped, there would be a particle that was at the same time in pure state w_2 and pure state w_2' or w_3'—which is impossible. Or, if we are not allowed to assign any states to individual particles, let us ask: what state belongs to the sub-ensemble which is the common part of the sub-ensemble with state w_2 and that with state w_2'? Shouldn't any sub-ensemble of an ensemble in a pure state, itself be in that pure state?

There is a way out of these difficulties if we content ourselves with a weakened form of the Ignorance Interpretation.[10] We can say that the specification of state w is *incomplete*, that there is information about nature that is lost in this assignment of a mixed state. Really, a system in state w is in one of the pure states that enter into some decomposition (like 1 or 2) of that state, and we don't know which. But the number $\frac{1}{3}$ by w_2 just tells us that the probability it is really in w_2 equals $\frac{1}{3}$ *if* 1 is the objectively right or true decomposition. And which is the true decomposition in this case is something we are not told in the quantum-mechanical description of nature.

To my mind, this is for many cases a gratuitously metaphysical addition; there is in general no physical difference to which it corresponds. It is perhaps a harmless addition, that will raise only academic eyebrows. However, even if we accept this, the Ignorance Interpretation runs into still further difficulties.

For there is a second situation in which mixtures naturally appear. This is in the results of interaction. Sometimes, after an interaction, two systems X and Y are again separated and mutually isolated, but we only have a pure state for the complex system. In such a case it may be inconsistent to ascribe in addition pure states to X and Y individually. (Schrödinger said that this was perhaps 'the' peculiarity of the new quantum theory.) In this case it is however possible to assign certain mixed states to X and Y ('reduction of the density matrix'), and indeed, this is generally done.[11]

But the Ignorance Interpretation renders this impossible. For according to it, ascription of a mixed state entails the assertion that the system is really in a pure state. So then, if we cannot consistently ascribe any pure state, neither can we consistently ascribe any mixed state.

What are those numbers $\frac{1}{3}$ then, those coefficients in the decomposition of a mixed state? They are called probabilities, and they can be interpreted as probabilities, but not in the naïve ways we have just tried. Before looking at the solution, let us examine the second problem.

The pure states are directly representable by means of vectors in a Hilbert space.[12] There are several such representations available; in the 'Schrödinger picture', such a vector is a time-dependent wave function ψ. It proved at first quite difficult to interpret exactly what this vector signified. Schrödinger himself began by suggesting it stood for a physical wave. The probabilistic interpretation was introduced by Max Born in his study of collision processes. We must distinguish carefully, however, between the basic equations relating the state ψ to certain probabilities, which are known as the *Born rules*, and the simple statistical interpretation which Born himself gave to these to begin. The former became the bridge that links quantum mechanics to the observable phenomena; the latter proved untenable.

The problem with which Born began was how electrons are scattered in collisions with atoms.[13] The pure state ψ of the combined system atom plus electron, after the collision, is a superposition of different states ψ_r corresponding to the different possible directions r in which the scattered electron may emerge. Born proposed the interpretation that $|\psi_r|^2$, that is the squared absolute value of the amplitude, is the probability that the electron is scattered in direction r. If we express ψ_r as a scalar multiple $c\varphi_r$ of a unit vector φ_r, then that probability is $|c|^2$; and this was the first occasion when the squares of the coefficients in the expansion of a pure state in terms of eigenstates of a physical quantity were interpreted as probabilities, in the way which is now standard.

Generalizing on this in a later paper, Born proposed that an electron, for example, has at all times a well-defined position and momentum. But the quantum-mechanical state gives only probabilistic information about the values of these (and other) physical quantities. Hence to ascribe a pure state is to make a complex probability assertion, in just the way that ascribing a macrostate is in classical physics. The difference would only be that in classical physics, it was the theory of the underlying microstates that was best understood, whereas in quantum physics, the laws governing the basic physical quantities (position, momentum) as opposed to the laws governing the probabilities about them (as summarized in the quantum-mechanical state) were totally unknown.

This interpretation which Born proposed for the probabilities he introduced, breaks down when confronted with the famous *double-slit* experiment.

There are so many expositions of this experiment (or, thought experiment) that I shall be very brief. To begin with, consider a television tube. A cathode emits a stream of electrons, which are passed through collimating slits so as to produce a beam of electrons. The electrons in this beam are deflected in various ways by a magnetic field so as to paint an utterly moronic picture on the screen. Suppose now that we eliminate the electromagnets and re-place them by a tungsten plate, inside the tube, which has two slits, that can be individually opened and closed. If each electron has at all times a well-defined position and momentum, then each electron that reaches the screen must have gone through one slit or the other. We can produce the electrons so slowly that they cannot interfere with each other. Consider now the little area X on the screen, and ask what the probability is that some electron hits X. That clearly depends on which slits are open;

   situation x: top slit is open only
   situation y: bottom slit is open only
   situation xy: both slits are open.

Let us call the corresponding probabilities $P_x(X)$, $P_y(X)$, $P_{xy}(X)$. Since the electrons cannot interfere with each other, we conclude:

$$P_{xy}(X) = cP_x(X) + dP_y(X) \text{ for numbers } c, d \text{ such that } c + d \leqslant 1.$$

That can be compared with the proportions of electrons that do fall on area X in these different situations; and it is found to be false. (If a similar experiment is done with an airgun and BB bullets, the formula checks out as true, of course.)

It is sometimes said that electrons are neither waves nor particles but wavicles, behaving now like one, then like the other. That is fine as a way to label the problem; but it does not constitute a solution to the puzzle of why that formula does not work. What is needed is an interpretation of the pure states that holds on to their association with probabilities, which Born proposed and which checks out wonderfully well in experiments, but does not imply that the probabilities in the double-slit experiment are related according to that addition formula.

§3.2 *Quantum Probabilities as Conditional*

The equations whereby Born related the state-vector to the probability that an electron is scattered along a certain direction, were generalized to other physical quantities (*observables*). Each physical quantity (position, momentum, spin, etc.) has a range of possible values. Without real loss of generality we can take these values to be numbers. For each measurable set E of numbers, each observable m, and each state w, the Born rules allow us to calculate a number between zero and one inclusive:

1. $P_w^m(E)$

which Born called the probability that observable m has a value that lies in E, for systems in state w.

For each state w, measurements of observable m on systems in that state yield results that fall inside set E with a proportion that closely accords with the quantity 1 calculated by Born's rules. Yet in view of the difficulties discussed, what Born called it, cannot be quite right. Hence Born's proposal was modified and led to the central tenet of the *Copenhagen interpretation*: This number is the *conditional probability* of a result in E *given that* observable m is measured on a system in state w. The revolutionary difference is that, on this interpretation there is no implication at all that the observable has any particular value, or indeed, any value at all, when no measurement is made. Let us see how this new interpretation solves the difficulties.

Let us begin with the double-slit experiment. The bright spot that appears on the screen indicates where the electron landed. So it is an indication of position, and we may regard the screen as a position-measurement apparatus. What state is the electron in when it is subjected to this measurement? Well, whatever state is prepared by the device consisting of cathode, collimating slits, and tungsten plate. That state-preparation device is of one of three sorts, depending on which slit(s) in the tungsten plate are open. Let us call the three prepared states

φ_x: top slit open only
φ_y: bottom slit open only
φ_{xy}: both slits open.

What does the theory have to say about these states? That φ_{xy} is a *superposition* of φ_x and φ_y. If we now calculate the probabilities of the position measurement giving us a result in X, *given that* such a measurement is performed on an electron in one of these states, we find a formula of general form

$$P_{\varphi_{xy}}^m(X) = e P_{\varphi_x}^m(X) + d P_{\varphi_y}^m(X) + f(x, y)$$

and this formula does correspond closely to the observed frequencies.

There is no contradiction at all here with probability theory, since we are conditionalizing on the same sort of measurement being made, but on electrons in different states. Comparing this with Born, we find the following difference. To say that φ_{xy} is a superposition of φ_x and φ_y, means that these vectors are related by an equation of the form

$$\varphi_{xy} = a\varphi_x + b\varphi_y$$

Now Born interpreted this as follows: $|a|^2$ is the probability that a system in state φ_{xy} is really in state φ_x, and that in addition means: it has a trajectory that goes through the top slit in the tungsten plate.

The Copenhagen interpretation is: $|a|^2$ is the probability that *if* we made a different sort of measurement on the electron, by placing a screen right behind the top slit of the tungsten plate, we would get a light spot. And indeed, if we do so, the frequency of such results does closely accord with $|a|^2$. But in the old Born interpretation this was taken to show that a corresponding fraction of the electrons goes through the top slit, even if this second sort of measurement is *not* made. In the Copenhagen interpretation, this is not assumed.

Turning now to the problem about mixtures, we get exactly the same story. To say that the system is in state

$$w = aw_1 + bw_2 + cw_3$$

indicates only a relation among those conditional probabilities about measurement, namely that

$$P_w^m(E) = a P_{w_1}^m(E) + b P_{w_2}^m(E) + c P_{w_3}^m(E)$$

for all observables m and all measurable sets E of values of m. There is no implication at all, in this formula relating conditional probabilities of measurement results, that the system is really in one of those pure states, nor any other implication about what is the case if no measurement is made.

Of course, this approach raises some serious questions. First of all, a measurement is itself a physical interaction, and hence, a process in the domain of applicability of quantum theory. So there is a serious consistency problem: does what quantum theory says about such processes cohere with the role they play in the Born rules linking states with measurement outcomes? This is called the *measurement problem* and is still a central topic of discussion in the philosophy of physics.[14] Secondly, if $P^m_w(E)$ is a conditional probability, then there must be a single probability function P^* such that, for all w and m and E,

$$P^m_w(E) = P^*(\text{outcome is in } E/\text{system is in state } w \text{ and subjected to an } m\text{-measurement})$$

where the traditional concepts of probability and conditional probability must apply. This is not at all obvious; but it can consistently be maintained.[15]

§3.3 *Virtual Ensembles of Measurements*

The Copenhagen interpretation restores some order to the subject. It allows us once more to view the probabilities as measures of objective quantities, namely frequencies of outcomes in sets or series of measurements. But there is a serious problem that remains.

In the classical case, the objective correlate of the probability was the proportion of time a system of the relevant sort would spend in a relevant sort of state. The quantum probabilities relate to proportions in classes of interactions of certain types (measurement inter-actions). What if no, or few, such interactions happen?

Of course, by calling some interaction a measurement we do not imply that a person 'performs' it. Nor need the 'apparatus' be something man-made. The general definition of a measurement interaction used in foundational studies has nothing to do with human agency or human history. So there are no doubt many more measurements, in the appropriate sense, than we may suspect. Indeed, there is no reason to think that there cannot be infinitely many, if the universe as a whole lasts for ever.

Still, neither do we want the success of the interpretation to rest on the empirical assertion that there are in fact infinitely many, or even a great many, measurements of *each* observable on systems in *every* state. In fact, that this is so, has zero probability. (The reason is that there are as many maximal observables, and also as many pure states, as there are real numbers (non-denumerably many). Each measurement takes a finite amount of time, with a definite lower limit on how short it can be. Together with the theorem of probability theory that if each of a class of mutually incompatible events has non-zero probability, then there are only denumerably many, this implies the conclusion I stated.) What has happened is that modality has raised its insistent head: the conditional probabilities seem to be about *what would happen if* ... And the class of events in which $P^m_w(E)$ is a proportion, is what is politely called a *virtual ensemble* of measurements. This means something like: an ensemble, i.e. a class, of events that do not happen, possible events.

Whatever complacency we had about the ease of a purely empirical interpretation of probabilities in physics, at any earlier stage, should by now have disappeared. We must turn, therefore, to the general problem of giving such an interpretation.

§4. *Towards an Empiricist Interpretation of Probability*[16]

Looking at physics, we saw probability in two roles, epistemic and objective—appearing sometimes as a measure of ignorance and sometimes as a measure of some objective feature such as proportional time of sojourn. The study of epistemic probability we can leave to epistemology proper. But probability as a measure of objective features of the world—or of features of the model that is meant to fit the world—must be dealt with in the philosophical analysis of the scientific description of nature.

As always, we shall have to answer two questions: what exactly does a probabilistic theory say? *and*, what do we believe if we accept this theory as empirically adequate? The two cannot be pursued separately, for to answer the second, what we must know is what the theory says about the observable phenomena; and that is part of the answer to the first. To put it in more neutral terminology, we need to understand the structure of the probabilistic models provided by our scientific theories, and also the way in which these models are meant to fit the data.

§4.1 *Probability Spaces as Models of Experiments*

There is a simple explanation of the objective probabilities that scientific theories give us, which is found in many texts. I shall begin with a short, modern account of probability theory, which will recapitulate the axioms mentioned previously, and which will proceed by means of that familiar explanation of what it is all about.

A *probability space S* consists of three parts: the sample space K, the family of events F, and the probability measure P. The sample space is also called the outcome-space, and the reasons for both terms lie in the intuitive interpretation. We envisage an experiment, whose possible outcomes are represented by the elements of K. One special sort of experiment consists in drawing a random sample from a certain population, in which case the outcomes which are possible are that your draw selects this or that item from that population. In all cases it is appropriate to call K the outcome-space, and in the latter, you can call it the sample space.

But actually, K is 'too fine' for the experimental report. Suppose I 'draw' an arbitrary cigarette smoker, and I inspect him for different types of cancer, emphysema, heart disease, and the like. The person I actually 'drew' was Claretta Playersplease; but that information does not go into the report. She was in perfect health in every respect—that goes into the report. The significant outcome-event was that the person selected was in perfect health.

This is the basic rationale for specifying a family F of 'experimentally significant' events. These correspond to subsets of K. For example, the set of perfectly healthy people is a subset of the cigarette smokers, and that the person selected was in that subset, is a (significant) outcome-event. To give oneself a picture, it serves to draw the sort of Venn diagram used in elementary logic. The set K is drawn as a square, and its elements correspond to the points in that square. Events in F are drawn on K as circles. If circle E is inside circle E', that means that whenever E happens E' must happen too. For example, you cannot select a mentholated Virginia cigarette smoker without, *a fortiori*, selecting a smoker of American cigarettes. If two circles overlap, their common part must also represent an event in F; this is called the *intersection* of the first two events. If $(E \cap E')$ is the intersection of E and E', then $(E \cap E')$ happens if and only if both E and E' do. Similarly, the region that encompasses both E and E', is an event $(E \cup E')$ called the *union* of E and E'; it happens if and only if either E happens or E' happens. Finally, the region which is the square minus the region E is called the *complement* $(K - E)$ of E, and is (the region corresponding to) an event which happens if and only if E does not happen.

A family of subsets of K which includes K, and is thus closed under the operations of intersection, union, and complement is called a *field*. We require that F is indeed a field. In the case of probability spaces used in physics, we require in addition that F is a *Borel field* (also called *sigma-field*) which means that it is closed under *countably infinite* unions:

if $E_1, \ldots, E_k, \ldots$, are in F, then
$\cup E_i = E_i \cup \ldots \cup E_k \cup \ldots$, is also in F.

This requirement makes a difference only if K is infinite.

Turning now finally to the probability measure P, this is a function $P(E) = r$ such that

   0. $P(E)$ is defined if and only if E is a member of the family F
   I. $P(K) = 1$; $0 \leqslant P(E) \leqslant 1$
   II. $P(E \cup E') = P(E) + P(E')$ if E and E' do not overlap (are *disjoint*)

To say that E and E' are *disjoint* means that they cannot possibly both happen. For example, the random draw of a cigarette smoker cannot produce someone who is perfectly healthy *and* has lung cancer. So E and E' are disjoint exactly if $(E \cap E')$ is the *empty* event, or *impossible* event, the one that cannot happen. For the probabilities used in physics we require furthermore

   II*. $P(\cup E_i) = \sum_{i=1}^{\infty} P(E_i)$ if the events

   $E_1, \ldots, E_k, \ldots$, are all mutually disjoint

A function satisfying II is called additive or finitely additive; if it satisfies II*, countably additive.

There is now also a theoretical reason for distinguishing the family F of 'significant' events. Why, after all, do we not just count every subset of K as a possible outcome-event, and just say that some of them are interesting and others, not interesting? The reason lies with the sort of probability spaces needed in physics. Let us take as sample space K just an open sphere in ordinary, three-dimensional Euclidean space, which could be the outcome-space for a position measurement. Suppose we have total uncertainty as to the outcome, and represent this by a measure P, for which we ask that it be defined for each subset E of points in K, that it satisfies I and II, and that if E and E' are congruent to each other they receive the same measure $P(E) = P(E')$. Hausdorff proved that no such function P exists. He also proved that in various geometrically still simpler cases, no such function P exists if we impose the stricter requirement of countable additivity.[17] These results are of course as important for the concept of volume as for probability. In any case, they explain why we always specify separately the outcome-space K, and the family F of outcome-events for which a probability is defined.

To continue the graphic illustration, imagine the Venn diagram (square K; circles E, E'; further regions $(E \cap E')$, $(E \cup E')$, $K - E$) with one kilogram of mud heaped on it. The mass of the mud lying on circle E, expressed as a fraction of a kilogram, is the number $P(E)$. There are limiting cases; for example, the mud evenly spread over K (then $P(E)$ is proportional to the area of E) and also the case in which all the mud is heaped on one point x in K (then $P(E)$ is one or zero depending on whether x is in E or not).

It will not have been unnoticed that this intuitive account of a probability space as a model of an experiment is thoroughly modal: the probability attaches to the *possible* outcomes of the experiment. Whether the model bears any relation to observable reality cannot become apparent if the experiment is only done once. The clue to the relation between this model and the world must lie in what it is meant to say about repetitions of the experiment. Intuitively: more probable outcome-events should happen more often.

Finally, we must insist that a probabilistic model need not be of a (humanly performable) experiment. Hence various writers speak rather of a *chance set-up*. This is like an experiment, except that it can happen spontaneously in nature: any state of affairs with possible outcomes will do. A person with a coin he is about to flip is one chance set-up; a limpet about to go on its daily food-hunt is another. I shall continue to discuss probability spaces as models of experiments, but it is easy to see how the discussion can be adapted to chance set-ups in general, *mutatis mutandis*.

§4.2 *The Strict Frequency Interpretation*

Hans Reichenbach advocated and defended the interpretation according to which probability *is* relative frequency.[18] The assertion $P(E) = r$ means simply that, in the class of all experiments of the sort in question, the outcome E occurs in a fraction r of all cases. Many of the objections philosophers have brought against this interpretation have little cogency. Any probability assertion must be understood with reference to a (tacitly specified) *sort* of experiment, or *class* of events, or class of chance set-ups (the *reference class*). For example, if I fall from a ladder, this is an event which belongs to the class of human falls from ladders, but also to the class of animal falls from heights below 50 feet; and the proportion of legs broken in the two classes is certainly different. But of course we cannot evaluate a probability model unless we know what it is meant to be modelling. If we understand Reichenbach's theory as attempting to provide a truth-conditional semantics for everyday probability talk, there is a problem of specifying the reference class; but if our interest lies in the interpretation of probabilities in science, I see no such problem.

Somewhat more worrisome is the fact that if $P(E)$ is a proportion in a finite class, then it must be a rational number. Hence if we say, for example, that $P(E)$ is the reciprocal of pi, or of the square root of 2, we must (according to Reichenbach) imply that the experiment is in fact repeated infinitely often. About this, Reichenbach can be perfectly hardheaded. Suppose the experiment (or chance set-up) occurred finitely many times. Then the assertion that $P(E)$ is the reciprocal of pi is false; but it could still be approximately correct. What harm done if we say, in such a case, that the number provided is merely an approximation? In all important cases, the reference class will be so large that the approximation will be very close—or if it is not very close, we have other, more immediate reasons for calling the assertion false.

The point will be clearer if we look at how the finite and infinite cases differ, and how they are related. The reference class may be finite or infinite. A theoretical assertion of probability must very often be understood as conditional on the reference class being 'large enough'. There will still be implications for frequencies in smaller classes, but these will be much more approximate, or less reliable. This sort of consideration automatically leads us to the infinite case as the only 'pure' case, the only one for which we can make precise statements. For imagine I say:

1. The proportion of A's in a class of B's will be $\frac{1}{2}$ provided the class of B's is large enough.

One attempt to make this precise yields

2. There is a size x such that, in any class of B's of size $\geqslant x$, the proportion of A's equals $\frac{1}{2}$.

This is self-contradictory, since either x or $x+1$ will be odd, and the number of A's is an integer. Clearly the size of the class of B's itself places a limitation on the exactness of the assertion. So we must say something like:

3. There is a size x such that for any $n \geqslant x$, if a class of B's has size n, then the proportion of A's therein equals $\frac{1}{2} \pm \frac{1}{n}$.

This entails that as the size of the class of B's approaches infinity, the proportion of A's tends to the limit $\frac{1}{2}$. True; what 3 says is stronger than this limit assertion. But just in the way that 3 is stronger, it is unacceptable: there is no reason to take 1 to rule out fluctuations around $\frac{1}{2}$ that are larger than $\frac{1}{n}$ in classes of size n, provided only the fluctuations become smaller and smaller as the size increases. Hence, although according to Reichenbach a probability statement is literally no more and no less than a statement of relative frequency in the *actual* reference class (however large or small it may be), the probability assertions of physical science should be taken as relating to ideally extended, infinite long runs, and applying only approximately to smaller cases.

One main problem will be how we can tease out implications for actual, finite reference classes. Should these be regarded as random samples from non-actual infinite series? If so, *which* non-actual (possible) infinite series? But this speculation and question already introduces an element of *modality* which is foreign to the strict extensionalist interpretation with which Reichenbach began.

Instead, let us concentrate here on the pure case of an actual long run. Perhaps in the 'basic' case of interactions among elementary particles, say, this infinity is in any case actual. But have we here a tenable interpretation of probability?

In the preceding section we inspected the structure of a probability space. Reichenbach now asks us to look at a different structure: a series $s = s_1, \ldots, s_k, \ldots$, of actual outcomes (a *long run*). Each of these must be an element of K of course, and may or may not belong to various events $E, E', \ldots$ In addition, there is the function of *relative frequency* which I shall call *relf*. We can define

4. $\#(E, s, n) =$ the proportion of outcomes in E which occur among the first n members $s_1, \ldots, s_n$ of s

5. $relf(E, s) = \underset{n \to \infty}{\text{limit}} \ \#(E, s, n)$

That is, *relf* (E, s), called the relative frequency of event E in the long run s, is the number which is the limit approached by the proportion of E among the first n outcomes, as we look at higher and higher numbers n.

This limit may not exist. To say that *relf* $(E, s) = r$, implies that the limit does exist, and equals that number.

The questions which this interpretation must now face are these: First, is it consistent to say that probability is the same as relative frequency? That means: have the functions $P(-)$ and *relf* $(-, s)$ the same properties? And secondly, even if it is consistent, is the interpretation not too narrow or too broad? That is, does relative frequency introduce structures like probability spaces which do not have the correct properties; or conversely, are some probability spaces simply impossible to relate to a relative frequency function for some long run? And the unfortunate fact is that Reichenbach's strict frequency interpretation does very badly with respect to all these questions.

Let the actual long run be counted in days: 1 (today), 2 (tomorrow), and so on. Let $A(n)$ be an event that happens only on the nth day. Then the limit of the relative frequency of the occurrence of $A(n)$ in the first $n + q$ days, as q goes to infinity, equals zero. The sum of all these zeros, for $n = 1, 2, 3, \ldots$ equals zero again. But the union of the events $A(n)$—that is, $A(1)$-or-$A(2)$-or-$A(3)$-or $\ldots$; symbolically $\cup \{A(n) : n \in N\}$—has relative frequency 1. It is an event that happens every day.

So relative frequency is not countably additive.[19] Indeed, its domain of definition is not closed under countable unions, and so is not a Borel field. For let B be an event whose relative frequency does not tend to a limit at all. Let the events $B(n)$ be as follows: $B(n) = A(n)$ if B happens on the nth day, while $B(n)$ is the impossible event if B does not happen on that day. The limit of the relative frequency of $B(n)$ exists, and equals zero, for each number n. But B is the union of the events $B(n)$, and the limit of its relative frequency does not exist.

A somewhat more complicated argument, due independently to de Finetti and to Rubin, and reported by Suppes, establishes that the domain of relative frequency in the long run is not a field either.[20]

Let me sum up the findings so far. The domain of definition of relative frequency is not closed under countable unions, nor under finite intersections. But still, countable additivity fares worse than finite additivity. For when the relative frequency of a countable union of disjoint events exists, it need not be the sum of the relative frequencies of those components. But if the relative frequencies of B, C, and $B \cup C$ all exist; while B is disjoint from C, then the relative frequency of $B \cup C$ is the sum of those of B and C.

We cannot say therefore that relative frequencies are probabilities. But we have not ruled out yet that all probabilities are relative frequencies (of specially selected families of events). For this question it is necessary to look at 'large' probability spaces; specifically, at geometric probabilities.

It is sometimes said that a finite or countable sample space is generally just a crude partition of reality—reality is continuous. Of course, you cannot have countable additivity in a sensible function on a countable space! But the problem infects continuous sample spaces as well. In the case of a mechanical system, we would like to correlate the probability of a state with the proportion of time the system will spend in that state in the limit, that is, in the full-ness of time. But take a particle travelling forever in a straight line. There will be a unique region $R(n)$ filled by the trajectory of the particle on day (n). The proportion of time spent in $R(n)$ tends to zero in the long run; but the particle always stays in the union of all the regions $R(n)$.

Geometric probabilities such as these have further problems. Take Reichenbach's favourite machine-gun example, shooting at a circular target. The probability that a given part of the target is hit in a given interval, is proportional to its area, we say. But idealize a bit: let the bullets be point-particles. One region with area zero is hit every time: the set of points *actually* hit in the long run. (This is because there are uncountably many points inside the circle, but only countably many points are hit.) Its complement, though of area equal to the whole target, is hit with relative frequency zero.

Because I gave rather a fanciful example, this conclusion may not seem damaging. But it is; it merely brings out in a graphic fashion that there are probability spaces which are not isomorphic to any such 'relative frequency space' resulting from a long run of outcomes with a relative frequency function *relf*. And those spaces—the geo-metric probabilities—are just the ones that are mainly used in physics.

Crucial to this reflection on the disparity between probability and relative frequency is the distinction between countable and uncountable infinity. There are finitely many positive odd numbers below 100; but there are infinitely many odd numbers. Yet these, since they are all integers, form only a countable infinity. The points on a line, or the real numbers between zero and one, already form an uncountable infinity—and these uncountable sets play a central role in the mathematical models of physics.

The laws of large numbers proved in probability theory are often mentioned as providing the link between probability and frequency. Provide a link they certainly do; make possible the strict frequency interpretation they do not. Without going into the technical details here, I can summarize the implications of the Strong Law of Large Numbers for the question whether Reichenbach could consistently maintain that probability is relative frequency:

6. If $E_1, \ldots, E_k, \ldots$, are countably many outcome events in probability space S (with K, F, P as before) then there is a long run $s = s_1, \ldots, s_m, \ldots$, such that $P(E_i) = relf(E_i, s)$ for all $i = 1, \ldots, k, \ldots$

In the case of most probabilities discerned in physics, however, the family F of outcome-events is not countable.

Richard von Mises also developed a frequency interpretation of probability (though I am not sure that it should be called a strict frequency interpretation) and he reports a result due to Polya, which strengthens statement 6 above.[21] Again cutting the story short, let us call $\{E_1, \ldots, E_k, \ldots\}$ a *countable partition* of space S if there are countably many of these outcome-events; they are all mutually disjoint; and they are exhaustive, that is, $(\cup E_i) = K$. The Borel field F^* which they are said to *generate* is the smallest Borel field to which they belong; although uncountable this is in general part, and not all, of the family F of all outcome-events. Polya's result implies then:

7. If F^* is the Borel field of outcome-events generated by countable partition $\{E_1, \ldots, E_k, \ldots\}$ of outcome-events in probability space S, then there is a long run $s = s_1, \ldots, s_m, \ldots$, such that $P(E) = relf(E, s)$ for all E in F^*.

We may conclude therefore that the identification of probability with relative frequency is logically consistent if it is restricted to specially restricted classes of events. This is important; but not sufficient to show us how the term *probability* is to be interpreted. In addition, even this restricted identification utilizes the relative frequency in an infinite sequence to be found in a model, and possibly not in the real world.

§4.3 *Propensity and Virtual Sequences*

In the nineteenth century, John Venn had already proposed the identification of probability with relative frequency. The American philosopher Charles Sanders Peirce had made the contrary suggestion that, although probability statements are about frequencies, they involve a 'would-be'—something like possibility or necessity or perhaps even 'many worlds' or ways of what might have been. Similarly, Reichenbach's strict frequency theory was followed by a blatantly modal 'propensity' theory.

In 1957, Karl Popper argued that probability is not a property of an actual course of events (such as experimental outcomes) but of the conditions in which these events occur (such as the experimental arrangement or chance set-up). He explicitly presented this as a more sophisticated version of the frequency interpretation, the difference lying in an appeal to what *would happen*, if the conditions were realized sufficiently often:

Every experimental arrangement is liable to produce, if we repeat the experiment very often, a sequence with frequencies which depend upon this particular experimental arrangement. These virtual frequencies may be called probabilities. But since the probabilities turn out to depend upon the experimental arrangement, they may be looked upon as *propensities of this arrangement*. *They characterize the disposition*, or *the propensity*, of the experimental arrangement to give rise to certain characteristic frequencies *when the experiment is often repeated*.[22]

We can put it this way: when we say that outcome A has a probability of $\frac{1}{2}$ in experiment C under present conditions, we mean that A *would* occur half the time, if experiment C *were* repeated sufficiently often. Thus we envisage a sequence $s: s_1, s_2, \ldots$ of outcomes, which is *the sequence* of outcomes *which would occur* if C were repeated indefinitely under the same conditions. The probability of A is the relative frequency of A in that sequence s; not its frequency in the actual course of events (which may end too soon, or in which the conditions of the experiment vary with time, etc.). Hence, Popper speaks of a 'virtual frequency'—the frequency in the 'virtual sequence'.

How is that probability assertion—i.e. the assertion about frequency in the virtual sequence—to be tested? Popper says explicitly that it is tested by looking at what happens in the actual sequence of events. But how can the actual sequence (which has, say, 1,000 members) provide a test of what the virtual frequency is? Only, as far as I can see, by being regarded as a random part of the virtual sequence. For we can begin with the premiss that in sequence s the relative frequency of A is $\frac{1}{2}$; then use a chi-squared table or similar test to see what we should expect to find in a random sample of 1,000 members—and compare this with what we find in the actual 1,000 outcomes.[23] The comparison would make no sense if we could not regard the actual sequence as a random selection from the virtual one.

Let us go a bit further. If the virtual sequence has been completely identified by what Popper says, then he must be saying that, if in the actual world the relevant conditions *do* occur infinitely many times (as nature does continually test our propositions!) then the actual frequency equals the virtual frequency.

I claim that he must be saying that; for otherwise he would be subject to the following objection. I could, for example, imagine doing the experiment once every minute, or once every two minutes, and so on. But when you say 'the sequence of outcomes that would occur' you do not specify the rate at which the experiments are performed. Hence your words do not pick out one definite frequency. The answer Popper must give, it seems to me, is that the sequence in question is performed at the rate at which it would be performed if we actually did it. Hence the words 'if C were ...', unlike the word 'could', pick out a single possible course of events which is determinate in all respects—and which is *indeed* realized if the conditions are realized. (How else could he speak of *the* virtual sequence?)

So now we know two things about Popper's virtual sequence: (a) the actual sequence is a random selection from it; and (b) if the actual sequence is infinite, the two are identical. This would seem to be the correct explication of Popper's rather vague account.

Popper called his new interpretation the *propensity interpretation*; subsequent advocates of this interpretation, however, do not all explain propensity as a frequency in what *would* happen under certain conditions. Some, like Ronald Giere, follow him only in saying that probability is an objective quality distinct from actual frequency, and refuse to say even that in an infinite sequence of experiments under correct fixed conditions, the relative frequency and the probability would coincide.[24]

Henry Kyburg does not advocate the propensity interpretation of probability. But in a provocative article, he outlined a way in which the propensity interpretation can be made precise.[25] To begin with, he notes the sorts of differences between Popper and other propensity advocates which I described at the end of the last paragraph. Some propensity theorists, like Giere, insist on a complete logical separation between probability and frequency (though asserting that beliefs about probabilities are the rational guide to expectations about relative frequencies). Others, like Popper, talk of probabilities as being the frequencies there would be under suitable conditions.

What Kyburg outlines is a view he calls the *hypothetical frequency interpretation*, which may satisfy propensity theorists of the second sort. The virtue of Kyburg's proposal is that it is stated in terms of mathematical models of experimental situations. I will explain it here in simplified form.

Suppose I measure a table leg three times, and find the lengths 100 cm, 100.3 cm, 100.5 cm. I assert that the real length is 100.3 cm ± 0.3 cm. This indicates my confidence that, if I were to keep repeating these measurements, the sequence of numbers obtained would have superior and inferior limits inside [100, 100.6]. There is no assertion that none of these measurements would yield values outside the interval, only that in the long run, those anomalous values would play a negligible role.

Experiments designed to test or establish probabilities are very similar. Suppose I toss a coin 100 times and obtain 49 heads. I say the coin is fair, indicating my confidence about a regularity that would be observed if this coin were to be tested by other people, say, tossing it respectively 10^3, 10^4, 10^5 times, and so on. So I am envisaging many other experiments which are not actually done. My model of the actual situation involves reference to many possible situations.

This way of modelling situations has been used with varying success elsewhere. In logic and linguistics, it has been used to develop hypotheses about such qualifiers as 'necessary', 'possible'; such adverbial modifiers as 'slowly', 'deliberately'; verbs such as 'believe', 'want', 'ought'; and tense and mood. In foundations of physics, the many-world view of Everett and De Witt is well known, if controversial, and such models have also been used for quantum logic, the measurement problem, and the quantum-mechanical paradoxes, and also for the axiomatization of classical and relativistic mechanics.[26]

The elements of a many-world model are referred to variously as 'set-ups', 'possible situations', 'possible worlds', or just 'worlds'. I shall use the last term, mainly for its brevity. In a Kyburg model, each world is or has a sequence of events in an outcome-space; most or all of these series are finite. A *maximal* world is one which is not part of another; or is infinite. In such a model, the probability of event *E* equals *r* if and only if, in every maximal world in the model, the relative frequency of *E* equals *r*. Sometimes there is no such number *r* in which case the probability of *E* is not defined.

This view differs from Popper's in that there is in general no single virtual sequence, but many; and the probability of an event equals *r* exactly if in all maximally extended series of possible experiments, the relative frequency equals *r*. This notion of possible is clearly not *logical possibility*; on the contrary, such assertions of possibility as are present here are themselves contingent, and are meant to reflect empirical facts. In this way we can similarly say that if a body were weighed twice in quick succession we would get approximately the same values for its weight; though from a *logical* point of view this is not necessarily so.

We can regard Popper's view as a special case: his models are Kyburg's, with the stipulation that the set of worlds has only one maximal element.

However, as they stand, both these views fall prey to all the difficulties that beset the strict frequency interpretation. For clearly, there are models containing but a single world (a single long run). In that case we find that the domain of the defined probability function may not be a Borel field, and indeed not even a field; and even where defined, not countably additive.

Given these difficulties (which some propensity theorists *can* and do avoid by the simple expedient of denying any and all logical connections between frequency and probability), the models will not be acceptable. But an improvement of Kyburg's proposal may be possible.

§4.4 *A Modal Frequency Interpretation*

A probability space is a model of a repeatable experiment or chance set-up. In some rigorously precise fashion, the probability function in that model must be linked with the frequencies of occurrences of outcomes. The attempts to make this linkage precise, which we have examined so far, were unsuccessful. But I think that we have now all the ingredients necessary for the representation of all (and only!) probabilities in terms of frequencies, as was required. To show this I shall go, step by step, through the idealizations that lead us from a description of an experiment to a probability space. This part of the discussion will have to include somewhat more attention to mathematical detail than before.

In an actual experiment we generally find numbers attaching to large but finite ensembles (whether of similar coexistent systems, or of the same system brought repeatedly into the same state) which we compare with theoretical expectation values.

But this comparison is based on the idea that the number found depends essentially on a (finite) frequency in just the way that the theoretical expectation value depends on a probability. Moreover, we demand closer agreement between the number found and the expectation value as the size of the ensemble is increased.

Hence it is reasonable to take as an *ideal* (*repeated*) *experiment* an experiment performed infinitely many times under identical conditions or on systems in identical states. The relation between the ideal and the actual should then be this: the actual experiment is thought about in terms of its possible extensions to ideal repeated experiments (its ideal extensions). If this is correct, we compare an actual experiment with a conceptual model, consisting in a family of ideal repeated experiments. In that model there should be an intimate relation between frequencies and probability so that the model can be directly compared with the theory under consideration. Secondly, the theory of statistical testing should be capable of being regarded as specifying the extent to which the actual 'fits' or 'approximates' that model of the experimental situation.

In any actual experiment we can only make finitely many discriminations. We can, for instance, determine whether a given spot appears with an x-coordinate, in a given frame of reference, of $y +$ 1 cm; with $-10^5 < y < 10^5$. The possible values of the x-coordinates are all real numbers, but we focus attention on a finite partition of this outcome-space. The *first idealization* is to allow that partition to be countable, that is, finite or countably infinite. Secondly, in the actual experiment we note down successively a finite number of spots; the *second idealization* is to allow this recorded sequence of outcomes also to be countable.

Note that theory directs the experimental report at least to the extent that the outcome-space (or sample space) is given; in different

experiments of the same sort we focus attention on different partitions of that space. So we begin with an outcome-space K and a Borel
field F of events on K, and describe one ideal experiment by means
of a countable partition $\{A_n\} \subseteq F$ and a countable sequence s of
members of K.

We also note that in the idealization, we could have proceeded
in different ways; for instance, we could have merely lifted the lower
and upper limits to y, or we could also have made the discrimination
accurate to within 1 mm as opposed to 1 cm. So our model will incorporate many ideal experiments with partitions $\{A_n^x\}$, all of which
are finer than the partition in the actual experiment. These represent
different experimental set-ups, and we cannot expect the series of
outcomes to be the same, partly because the change in the set-up
may affect the outcomes, and partly because we must allow for
chance in the outcomes: if many ideal experiments *were* performed,
just as in the case of actual ones, we would expect a spread in the
series of outcomes.

But we are building a model here of what would happen in ideal
experiments, and this model-building is guided either by a theory
we assume, or by the expectations we have formed after learning
the results of actual experiments. Hence we expect not only a spread
in the outcomes, but also a certain agreement, coherence. In actu-
ality, such agreement would again be approximate, but here comes
the *third idealization*: we assume that the agreement will be exact.

To make this precise, let us call *significant events* in an ideal experiment with partition $\{A_n\} = G$ exactly the members of the Borel field
$BG \subseteq F$ generated by G. We now *stipulate first* that, if that ideal experiment has outcome sequence s, then $relf(A_n, s)$ is well defined
for each A_n in G and that $\sum_{n=1}^{x} relf(A_n, s) = 1$. By Polya's result it follows
that $relf(-, s)$ is well defined and countably additive on BG. Only
the relative frequencies of those significant events generated by the
partition will be considered in the use or appraisal of any ideal experiment. We *stipulate secondly* that if A is a significant event in
several ideal experiments, or the countable union of significant
events in other such experiments, then the frequencies agree as required. To answer the question how many ideal experiments the
model should contain, we *stipulate thirdly* that the significant events
together form a Borel field on the set K of possible outcomes. This
must include, for instance, that if we consider experiments with parti-

tions $\{A_n^m\}$, $m = 1, 2, \ldots$, and the series A_1^1, A_1^2, A_1^3, $\ldots$, converges to event A, then there must be some experiment in the model in which A is a significant event.

Together, these three stipulations give the exact content of the third idealization. And the three idealizations together yield the notion of a model of an experimental situation by means of a family of ideal experiments ('*a good family*'), which I shall now state precisely.[27]

1. *A good family* (*of ideal experiments*) is a couple $Q = \langle K, E \rangle$ in which K is a non-empty set (the 'possible outcomes' or 'worlds') and E is a set of couples $\alpha = \langle G_\alpha, s_\alpha \rangle$ (the 'possible experiments') such that
   (i) G_α is a countable partition of K and s_α a countable sequence of members of K (the 'outcome sequence' of experiment α);
   (ii) if A_1, A_2, $\ldots$ are in BG_{α_1}, BG_{α_2}, $\ldots$, the Borel fields generated by G_{α_1}, G_{α_2}, $\ldots$ (the Borel fields of 'significant events' of α_1, α_2, $\ldots$) then there is an experiment β in E such that $B = \cup A_i$ is in BG_β; and $relf(B, s_\beta) = \sum \{relf(A_i, \alpha_i): i = 1, 2, 3, \ldots\}$ if the A_i are disjoint;
   (iii) $relf(A, s_\alpha)$ is well defined for each A in G_α;
   (iv) $\sum \{relf(A, s_\alpha): A \in G_\alpha\} = 1$;
   (v) if $A \in BG_\alpha \cap BG_\beta$ then $relf(A, s_\alpha) = relf(A, s_\beta)$.

As noted, in the experiment $\langle G_\alpha, s_\alpha \rangle$ we call s_α the *outcome sequence* and the members of BG_α the *significant events*. Polya's result guarantees, from (iii) and (iv) that $relf(A, s_\alpha)$ is defined for each significant event A. It will be noted from condition (ii) that the union F of all the BG_α is itself a Borel field, for, if A_1, A_2, $\ldots$ are significant events of α_1, α_2, $\ldots$, their union is in BG_β and hence in F.

2. If $Q = \langle K, E \rangle$ is a good family, we define, for A in F: $PQ(A) = r$ if and only if $relf(A, s_\alpha) = r$ for all α in E such that A is a significant event in α.

3. *Result I:* If $Q = \langle K, E \rangle$ is a good family with family F of significant events, then $\langle K, F, PQ \rangle$ is a probability space.

For let A_1, A_2, $\ldots$ be disjoint members of F. Then by condition (ii) of (1), there is an experiment β such that their union is a significant event in β. Their relative frequencies are the same in all experiments in which they are significant events, hence PQ is countably additive.

4. *Result II:* If $\langle K, F, P \rangle$ is a probability space, then there is a good family $Q = \langle K, E \rangle$ such that $P = PQ$.

This is easy to prove: let E be the family of all couples $\langle G, s \rangle$ such that $G \subseteq F$ is a countable measurable partition of K and s a sequence of members of K such that *relf* $(-, s)$ restricted to G is exactly P restricted to G. Condition (ii) of 1 is satisfied simply because the measurable sets are closed under countable union, and every measurable set is a member of some countable measurable partition. The other conditions are satisfied because P is a probability measure.

We now have the desired representation result: probability spaces bear a natural correspondence to good families of ideal experiments. It is possible therefore to say:

5. The probability of event A equals the relative frequency with which it would occur, were a suitably designed experiment performed often enough under suitable conditions.

This, in slogan formulation, is the modal frequency interpretation of probability.

§4.5 *Empirical Adequacy of Statistical Theories*

My main concern in the preceding sections was to elucidate what is said by a theory which implies assertions of probability. I have tried to show that if we look into a probabilistic model, we see a picture of various different configurations in outcome sequences in an infinitely repeated experiment. These outcome sequences are different one from another, but certain features are common to all; and it is these common features that determine the probability function in the model. Hence, to put the matter concisely, the assertion that *this* model is the correct one, means that the actual series of experimental outcomes will display these common features.

However, we are brought up short by the reflection that the actual outcome series may be finite. I believe that the fit of the model to the world in that case is to be regarded in exactly the same way as the fit of the model to the data obtained on such experiments *so far*. These data, clearly, relate to a finite series of outcomes, whether or not the actual long run of outcomes is finite. And the measurement of how well a probabilistic model fits the data gathered—that is a subject already of extensive study in statistics. I have been making remarks about this along the way (if the actual outcome series is

finite, it should be regarded as a finite random sample from the infinite series in the model—which, I should add now, can be the outcome series in any of the ideal experiments in which events actually
verified in these actual experiments are significant). Other philosophers have given more detailed attention to these measures of the
goodness of fit of a probabilistic model. I mention especially Patrick
Suppes; and to give the flavour, quote a representative passage. Discussing an experimental test of a linear-learning model conducted
by himself and others, he specifies how an estimated learning parameter yields a precise probability function for success in trials of
a certain sort. A series of such trials was held: each subject was run
for 200 trials; there were 30 subjects (undergraduate students at
Stanford University). The method of predicting frequencies of success in the actual outcome sequences so obtained, on the basis of
the probabilities in the model, is by standard statistical calculation.
These predicted frequencies are then compared with observed frequencies. At this point one can discuss how good a fit there is
between model and data:

The χ^2 test for goodness of fit between the predicted and observed values
yields a χ^2 of 3.49. There are four degrees of freedom, but one parameter
has been estimated from the data, and so the χ^2 is to be interpreted with the
degrees of freedom equal to 3 and, as would be expected from inspection of
Table 1, there is not a statistically significant difference between the predicted
and the observed data.[28]

To get an intuitive idea of how the theoretically predicted frequencies
are derived, imagine that you are told that 25 marbles are drawn
from a barrel containing one million marbles. On the hypothesis that
half are black, how many black ones do you predict that there will
be among those 25? Obviously, you want to do more than say: about
half. Your hypothesis is not tested unless we make more than one
such draw—that is why the above experiment used 30 subjects. The
chi-squared test mentioned in the above passage is a statistical
method for telling how well the results bear out the hypothesis. This
test can be used for probabilities, and not merely for finite barrels,
because its accuracy is not too sensitive to the size of the barrel, ever
the less so the larger the barrel is. Hence we can transpose this
measure of goodness of fit to the comparison of relative frequencies
in the long run with the actual frequency in the actual run, viewing
that actual run as a random selection from the model's 'barrel'.

Pending more detailed investigation, then, we may say that a statistical theory is empirically adequate if it has at least one model such that the difference between predicted and actual frequencies in the observable phenomena is not a statistically significant difference.

§5. *Modality: Philosophical Retrenchment*

In an earlier chapter, I argued that certain issues in philosophy of science (having to do with observation and the definition of a theory's empirical import) had been misconstrued as issues in philosophy of logic and of language. With respect to modality, I hold the exact opposite: important philosophical problems concerning language have been misconstrued as relating to the content of science and the structure of the world. This is not at all new, but is the traditional nominalist line, which was also followed in the modern empiricists' attempts to disentangle relations among facts from relations among ideas. To substantiate such a view requires, of course, a theory of language as well as a theory of science; and I shall not pursue the issue here very far.

§5.1 *Empiricism and Modality*

The models of a probabilistic theory are, or have parts which are, probability spaces. In the interpretation I proposed, each such model is reconstructed as consisting of elements each of which represents an alternative possible sequence of outcome-events. At most one of these sequences can correspond to the sequence of events (such as experimental outcomes) which actually happens in our world.

The arguments leading up to that interpretation were designed to show that we can have no more economical reconstruction of what goes on in a physical theory which involves probabilities. A reconstruction of a model of such a theory, in which every part corresponds to something actual, cannot be had.

If we are convinced of this, we shall also show less resistance to similar assertions about the models of other theories. If I model the behaviour of a simple oscillator, or a pendulum bob, I use a state-space (the phase-space) and describe the entity's trajectory in that space. Many points therein do not correspond to states the entity actually had at any time, and many trajectories in that space, allowed by our physics, bear no correspondence to the entity's history, nor to the actual history of any similar entity.

Such a space, and the model as a whole if it involves more, is a mathematical entity. Does it, as a whole, correspond to a real part of the world? Do the important substructures correspond to something real: to non-actual, but real states, real trajectories; or perhaps to the states or trajectories of real but non-actual entities; or to the states or trajectory which this real entity has in a different, real, but non-actual, possible world? Do the sequences of events in (my reconstruction of) a probability space correspond each to what happens in some real, but perhaps non-actual, situation?

It may be unfair to pose the questions this way. An affirmative answer to some of these, or similar questions is given by various philosophical positions; and it may be unfair to pose the questions without going into the reasons for such a position. What I do wish to do, however, is to show why, within the philosophical position I have developed here, my insistence on the modal frequency interpretation of probability (and, more generally, on the state–space approach to the foundations of physics) does not commit me to that sort of metaphysical position.

From my point of view, modal realism seems very similar to theoretical entity realism. If we look at a model of a scientific theory, we discern important substructures which do not correspond to anything observable; and we also see substructures that do not correspond to anything actual. The two cases overlap; there is no logical relation between observability and actual existence. (The ride of the headless horseman is an observable event, but not an actual one.) Philosophical attitudes towards the two may differ. But according to constructive empiricism, the only belief involved in accepting a scientific theory is belief that it is empirically adequate: all that is *both* actual *and* observable finds a place in some model of the theory. So as far as empirical adequacy is concerned, the theory would be just as good if there existed nothing at all that was either unobservable or not actual. Acceptance of the theory does not commit us to belief in the reality of either sort of thing.

I would still identify truth of a theory with the condition that there is an exact correspondence between reality and one of its models. This would imply that, if such a model has parts corresponding to alternative courses of events (alternative in the sense of mutually incompatible), then there can be a complete correspondence between the model and reality only if alternative possible courses of events are real. And logical relations among theories and propositions continue of course to be defined in terms of truth: the theory implies a proposition exactly if that proposition is true under any conditions under which the theory is true. But all this is parallel to what I have said about the unobservable: empirical adequacy does not require truth; in my view, science aims only at empirical adequacy and anything beyond that is not relevant to its success.

Let me introduce three assertions which are at first blush inconsistent with each other. *First:* probability is a modality. *Second:* science includes irreducibly probabilistic theories. And *third:* there is no modality in the scientific description of the world.

The preceding paragraphs will have explained why, and in what sense, I assert the second and third. But there remains the first. Probability is a modality; it is a kind of graded possibility. How shall we make sense of that; and what is there, really, to be made sense of? The answer should be part of the solution to a larger problem: to do justice to the appearance of modality in science.

In my opinion, that solution consists mainly in the correct diagnosis of the problem, which is that modality appears in science only in that the language naturally used once a theory has been accepted, is modal language. This relocates the problem in philosophy of language, for it becomes the problem of explicating the use and structure of modal language. So if anyone asks: 'What more is there to look at in science besides the models, the actual phenomena, and the relationships between them?' we can answer 'The structure of the language used in a context where a scientific theory has been accepted.' And the problem of doing justice to modality will have been solved to an empiricist's satisfaction if we can explicate the use and structure of that language without concluding that anyone who does use it is committed to some sort of metaphysical beliefs such as that alternative possible worlds are real.

§5.2 *The Language of Science*

In a historical context in which a certain scientific theory has been accepted, a certain form of language is naturally adopted. We do not say, the events of spontaneous decay of radium atoms are represented by such and such elements in this model, and there is a probability function defined on those elements. We say instead: the probability of spontaneous decay of a radium atom is thus and so. We do not say that burning of copper at room temperature and pressure has no counterpart in any model of our physics; we say simply that it is impossible. Once the theory is accepted (whether *tout court* or only momentarily or for the sake of argument) it guides our language use in a certain way. The language we speak at that point has a logical structure which derives from the theories we accept.

It must be remembered here what tasks a theory of language sets itself. It must account for the relevant phenomena; and these are mainly the grammatical structure and inference patterns exhibited in verbal behaviour. The models we construct, in developing such a theory, are artificial languages. If language use is guided by an accepted scientific theory, then we must look to that theory in order to construct models of the language in use.

I cannot pretend that we have a theory of language which is satisfactory or anywhere near complete. Nor can I embark on this major project here. But I will briefly explain what I regard as the two central problems for the explication of the language of science, and some ideas concerning their solution. It will already be clear, first, that I regard this as a subsidiary task (not contributing directly to the main tasks of philosophy of science) and second, that it is part and parcel of the task of developing a general theory of language. The language of science is a part of natural language and not essentially different from other parts.

The first main problem is this. In connection with the development of modal logic and its proliferating branches, and of recent theoretical linguistics, we have seen the development of a very rich formal semantics. There, a language is characterized (this is one way to look at it) by specifying the structure of models for theories formulated in that language. These are generally referred to as *possible world model structures*. On the other hand, in philosophy of science much attention has been given to the characterization of the structure of models as they appear in the scientific literature. The first central problem is to bring these two efforts together, because at first sight, the model structures found in semantics, and the models of scientific theories (even as found in the foundations of physics) are totally different.

What we should try to do here is characterize (fragments of) scientific language by means of the concepts of formal semantics but in such a way that the model structures derive in an obvious way from the models of scientific theories. There is a good deal of work that contributes to this, mostly in connection with the foundations of classical and quantum mechanics.[29]

The second central problem belongs to the development of

pragmatics. We need to model linguistic situations in a way that reflects how language use is guided by assumptions, suppositions, accepted theories (equivalently: how the structure of the language used is determined by suppositions and theories which have been accepted). This second problem may be solved to some extent, if we have a good solution to the first. For that would mean that the family of model structures of a language (that is, the family of structures which are models for theories formulated in that language) is derived from the family of models which is provided by (or constitutes) a scientific theory. This would give us automatically some insight into how changes in theories accepted precipitate changes in the structure of language used. But there is more to it. We need to develop the peculiarly pragmatic concepts that are applicable here. If you and I, in dialogue, make a supposition, then it is thereafter *correct to assert* what follows from that supposition, and incorrect to assert what is inconsistent with it. Many other things remain correct to assert, because they are true or because we obtain evidence for them. Yet some things may be true and incompatible with the supposition. Since this is recognized by the discussants, it is not accurate to say that they speak just as if they believe the supposition to be true. The development of the concepts and methods needed for a formal pragmatics is now a lively research area. I have already drawn on its early successes to explicate the context-dependence of why-questions. But for the crucial question of how supposition and theory acceptance guide language use, we have only a few preliminary studies.[30]

Let us close with an example of how the logical structure of a language can be determined by an accepted theory: Wittgenstein's familiar example of the colour spectrum as a 'logical space'. A person uses a language in which he asserts such sentences as

1. X is green, X is not red, Y is red, ...
2. Nothing that is green is red
3. There is no possible object which is both green and red.

Sentences of type 1 he has been trained or conditioned to assert under certain experiential conditions. Sentences of type 2 still express assertions which are merely about what is actually the case. But 3 goes well beyond that; it says something like: there *could not be* something which is both red and green.

The explanation is that this person is guided by his idea of a simple abstract structure, the colour spectrum. We can think of this as a line segment, or an interval of real numbers (the wave lengths). He associates with each colour predicate, such as 'green', a part of that spectrum; he associates disjoint parts with 'red' and 'green'; and when he says that an object is green or red, he is *classifying it*, that is, assigning it a *location* in that spectrum. So sentence 2 amounts to: no *occupied* location belongs at once to the parts associated with 'red' and with 'green', while 3 says that no point of the spectrum at all belongs to both parts. (Every model structure of this simple language consists of that spectrum plus a domain of entities and a function that assigns a location in the spectrum to each of those entities.) It will be clear how the logical connections among sentences in this language are determined by the structure of the colour spectrum. Blatantly modal sentences (such as 3) do occur; but this person evaluates them as true or false by reflection on the structure of the spectrum that guides all his uses of colour terms. His linguistic commitments can be summed up by referring to his use of this spectrum; his theory of colour consists in the family of models each of which is a classification of objects through location in this spectrum.[31]

§5.3 *Modality without Metaphysics*

Concern with the language of science is a typically philosophical preoccupation. It has certainly helped at times to throw light on philosophical problems concerning the structure of scientific theories, and concerning the relations of those theories to the world. But the help has always been indirect, and it has misled and hindered as often as helped. As philosophers we must try to separate problems concerning language from problems concerning science specifically; but also, as philosophers, we cannot limit ourselves to the problems peculiar to any one subject.

Guided by the scientific theories we accept, we freely use modal locutions in our language. Some are easily explicated: if I say that it is impossible to observe a muon directly, or to melt gold at room temperature, this is because no counterpart to such events can be found in any model of the science I accept. But our language is much subtler and richer than that; its modal locutions reflect the fact that in the models of our theories we see structures that correspond to alternative courses of events, not all of which could be jointly actualized.

On the view of acceptance of theories which I have advocated under the name of constructive empiricism, it does not require belief that all significant aspects of the models have corresponding counter-parts in reality. This applies to many aspects discussed by philos-ophers of science: space-time, elementary particles, fields, and, fin-ally, alternative possible states and courses of events. The locus of possibility is the model, not a reality behind the phenomena.

At the same time, acceptance has a pragmatic dimension: it in-volves a commitment to confront any phenomena within the con-ceptual framework of the theory. A main way in which this shows itself is that the language we talk has its structure determined by the major theories we accept. That is why, to some extent, adherents of a theory must talk just as if they believed it to be true. It is also why breakdown of a long-entrenched, accepted theory is said to pre-cipitate a conceptual breakdown, and why it is natural to speak of conceptual revolutions. For in theory change, the logical structure of our language in use may change. However, we are much more flexible in language use than many philosophers seem to assume: we are quite used to suspensions of belief or of conceptual com-mitment in dialogue with adherents of theories which we personally do not accept. This prepares us for such eventualities—it is note-worthy that radical as scientific revolutions have been, and confused as concepts and language sometimes became, scientists never became tongue-tied, but always successfully (if gradually) adapted their lan-guage to the changing tides of theory.

Since all men are mortal, commitment to a theory involves high stakes. The theories we develop are never complete, so that even if two of them are empirically equivalent, they will be accompanied by research programmes which are generally very different. Vindica-tion of a research programme within the relatively short run may depend more on the theory's conceptual resources and facts about our present circumstances than on the theory's empirical adequacy or even truth. That is why the commitment involved in acceptance of a theory runs so deep, and why we need not postulate belief in its truth to account for that. The depth of commitment is reflected, just as in the case of ideological commitment, in how the person is ready to answer questions *ex cathedra*, using counterfactual con-ditionals and other modal locutions, and to assume the office of explainer. Scientific realists and empiricists are equally in need of a deeper analysis of these aspects of scientific activity.

To be an empiricist is to withhold belief in anything that goes beyond the actual, observable phenomena, and to recognize no objective modality in nature. To develop an empiricist account of science is to depict it as involving a search for truth only about the empirical world, about what is actual and observable. Since scientific activity is an enormously rich and complex cultural phenomenon, this account of science must be accompanied by auxiliary theories about scientific explanation, conceptual commitment, modal language, and much else. But it must involve throughout a resolute rejection of the demand for an explanation of the regularities in the observable course of nature, by means of truths concerning a reality beyond what is actual and observable, as a demand which plays no role in the scientific enterprise.

7

Gentle Polemics[1]

It is only shallow people who do not judge by appearances. The true mystery of the world is the visible, not the invisible.

Oscar Wilde, *The Picture of Dorian Gray*

It is always easy to tell whether people are doing good philosophy: they are if they are laughing.

Charles Daniels

NOT long ago, I refused to believe in the existence of the theoretical entities postulated by science. I agreed, of course, that science postulates subatomic particles, forces, fields, and what have you, in order to describe the regularities found in nature. And I admitted readily that there are regularities, and that science rightly aims at a unified and coherent account of this immense—surprising or boring—regularity in phenomena. However, I regarded the theoretical entities as fictions facilitating systematic account, and not as providing a true explanation.

But now I must report on the reasons which have converted me to total belief in scientific realism.[2] This change of mind was a sudden occurrence, taking me unawares when I was reading Aquinas. Like Saul on the road to Damascus, I was struck by a blinding light, and I saw. What I saw was that the medieval attempts to prove the existence of God have modern analogues demonstrating the correctness of scientific realism.

Indeed, exactly where the invalidity of the proofs of God's existence is most obvious, the truth of scientific realism veritably springs to the eye. Paley described a watch found, running perfectly and keeping time, on a deserted heath.[3] Can you conceive, he asked, of a watch without a watchmaker? And the twentieth-century reader is blasé enough to answer: Yes, I can. But think again, gentle reader: can you conceive of a watch keeping time *without clockworks inside*?

In the *Summa Theologiae* and the *Summa contra Gentiles*, Aquinas set forth the Five Ways in which, in his opinion, God's existence can be proved. I shall present an analogue to each of the Five Ways to demonstrate the correctness of scientific realism.

I

I shall begin by stating Aquinas's First Way (roughly following the *Summa contra Gentiles* I, 13). Since my present topic has nothing to do with the existence of God, I shall keep my commentary on the original 'proofs' as short as possible.

Thus Aquinas: Everything that is moved, is moved by another. That some things are in motion is evident from the senses: for example, the sun. So we must either proceed to infinity, or arrive at some unmoved mover. But it is not possible (in this) to proceed to infinity. Hence there is an unmoved mover.

And Aquinas continues: In this proof, two propositions themselves need proof; namely, that whatever is moved, is moved by another, *and* that in movers and things moved, one cannot proceed to infinity.

In commentary on Aquinas it has often been pointed out, *one*, that he is not restricting motion here to 'local motion' but is thinking of any kind of change, such as heating or becoming wet; and *two*, that he need not be understood as denying the possibility of an infinite past. Indeed, he says quite explicitly that the fact that the world has a beginning in time cannot be demonstrated. It is rather a regress in causation of movement or change that he denies; so if the causal order has no beginning, then it as a whole needs to have a cause. As Paley said about his watch: if we found that it included a mechanism for the production of further watches, and hence was likely to be the offspring of a long line of heath-dwelling clocks, this would only increase our admiration for the genius of the watchmaker.

All this strikes us, nevertheless, as rather naïve, for we have learned to live with a great deal of scepticism, not only for the need to postulate causes but with respect to the very notion of causation. However, when we consider the reality of the theoretical entities of science, we see that their relation to the order of nature is a much more sophisticated one. For their existence is postulated to *explain* the regularities in nature. And if causation is one of the less respected and less studied notions in contemporary philosophy of science, explanation is a subject of the liveliest interest.

So I argue: Everything that is to be explained, is to be explained by something else. That some things are to be explained is evident, for the regularities in natural phenomena are obvious to the senses and surprising to the intellect. So we must either proceed to infinity, or arrive at something which explains, but is not itself, a regularity in the natural phenomena. However, in this we cannot proceed to infinity.

And I continue: In this proof, two propositions themselves need proof; namely, that whatever is to be explained, is to be explained by something else, *and* that in explanation one cannot proceed to infinity.

With respect to the former, I can appeal to the discussion of explanation initiated by the famous paper of Hempel and Oppenheim, and lasting to this very day.[4] It is a necessary condition for the explanation of conclusion C by premisses A_1, ..., A_n that A_1, ..., A_n deductively entail C. (Though in normal usage an explanation may be elliptic; that is, some premisses may not be explicitly supplied.) But it is also a necessary condition that A_1, ..., A_n entail more than C. For the fact that this bar attracts iron filings is not explained by noting that it does so; it is explained by the conjunction of the specific premiss that this bar is a magnet, and the generalization that all magnets attract iron.

Now this opens up the possibility, prima facie, that regularities initially noted may be explained by more comprehensive regularities, still on the phenomenal level, and so on *ad infinitum*. However, an infinite series of ever more comprehensive regularities in natural phenomena is at least as much in need of explanation as a single one. Indeed, more so. This is made very clear by J. J. C. Smart in his discussion of Craig's theorem. Craig would eliminate theoretical terms and replace a scientific theory T by a description of an infinitely complex regularity, entailing exactly the observational consequences of the original theory. The original theory T is finitely axiomatized, but its Craigian transform T' has infinitely many axioms, catalogued metalinguistically:

And if T is just a lot of ink marks, and is not capable of an objective interpretation, it does seem too much of an unbelievable coincidence that T′ should work at all. It is beyond my powers of credulity. ... The fact that in constructing T′ we do not need to mention the entities mentioned in T but only their names, does not remove the feeling that there would have to be an infinite number of coincidences if T were not capable of interpretation and objectively true.[5]

In other words, what would purport to be an infinite regress in explanation, would amount to no more than the postulation of an infinite coincidence. And that is just too much of a coincidence.

Aquinas was possibly not entirely happy with his argument that in movers and things moved, one cannot proceed to infinity. For it seems reasonable to me to read his Second Way as intending to reinforce exactly this part of the argument in the First Way. My analogue of the Second Way will play exactly that role with respect to this part (about infinite regresses in explanation) of the foregoing argument.

II

The Second Way concerns efficient causation, and I see it as a substantial contribution to the First Way, though Kenny regards it only as a report on medieval astrology.[6] From here on, I shall follow the exposition in the *Summa Theologiae* (*Qu.* 2 *Art.* 3).

Thus Aquinas: In the world of sensible things, we discern an order of efficient causes. There is no case of a thing being its own efficient cause; for if so, it would be prior to itself, which is impossible. Nor is it possible to proceed to infinity with efficient causes. Therefore it is necessary to admit a first efficient cause.

And Aquinas supports this: In the causal order, the first is the cause of the intermediate cause, and the intermediate is the cause of the ultimate cause. Now to take away the cause is to take away the effect. Therefore, if there be no first among efficient causes, there will be no ultimate, nor any intermediate cause; and hence no ultimate effect.

It must be pointed out again that 'prior' (*prius*) need not be interpreted temporally. To state my analogue for explanation, I appeal to a point made in I: for A to explain B it is a necessary condition that A entail B; and also that A entail more than just B. Furthermore, as the Hempel account also specifies, the premisses of an explanation must be true. (Note that I appeal to the Hempel account only as establishing *necessary* conditions for explanation, thus avoiding the major putative objections to that account.) I shall accordingly say that A *properly entails* B if and only if *one*, A is true; and *two*, A entails B, and *three*, A entails more than B.

So I argue: In explanation of natural phenomena, we discern an order of proper entailment. It is impossible for something properly to entail itself, for it cannot entail more than itself. Nor is it possible to proceed to infinity with proper entailments in an explanation.

And I support this: In an explanation, the premisses establish the truth of the intermediate lines, which establish the truth of the conclusion. But if there be no premisses (lines not properly entailed by preceding lines, singly or in conjunction), then no truth is established at all.

And I demonstrate this by supposing the contrary *per absurdum*.[7] Suppose, that is, that in the usual definition of *proof* (a finite sequence of lines each of which is an axiom or entailed by preceding lines, singly or in conjunction), we omit the restriction to a *finite* length. Then there might be infinitely long derivations of facts from theories, which could be offered as explanations of these facts. However, this is absurd since every true statement can then be 'explained' if any can. For imagine that there exists an *indispensable* infinite explanation: an infinite series of statements $\ldots, B_n, \ldots B_1$ such that each is properly entailed by its predecessor, and no true statement entails all of them already. Then let A be any true statement whatsoever, and construct the series

$$\ldots, B_n \mathbin{\&} A, \ldots, B_k \mathbin{\&} A, A$$

where k is the first number such that A does not deductively entail B_k. It is clear that in this new sentence, if A is any true statement whatsoever, then each line is properly entailed by the preceding line. Indeed, this sequence is—the restriction of finitude for proofs, and also the truth-value of A, left aside—an unimpeachable, mathematically correct, *categorical* proof of the statement A. For each line is either a logical axiom or follows deductively from preceding lines. But this reduces the subject to absurdity.

The import of this is to support the crucial lemma about the inadmissibility of an infinite regress in explanation, which appeared in I.

III

The Third Way is the proof *de contingentia mundi*. It has the apparent drawback of involving an obvious logical fallacy. It is again in the spirit of St. Thomas's enterprise to eliminate its apparent reference to time; and this removes the fallacy.

Thus Aquinas: The Third Way is taken from possibility and necessity, and runs thus. We find in nature, things that are possible to be and not to be. But it is impossible for these always to exist, for that which can not be, at some time is not. Hence, if everything were thus, there would have been a time when there was nothing. But if this were true, even now there would be nothing.

Reinterpreted, this Way is concerned with the *intelligibility* of the world.[8] That some particular fact is thus or so, is contingent; but it may be explained by pointing out that it is contingent on, and a consequence of, the world being thus or so. However, that the world is thus or so, is contingent too; and the immediate question is: contingent on what?

Be that as it may (I hold no brief for the soundness of these proofs as proofs of God's existence), there is certainly much relevance to scientific realism in a consideration of natural necessity.

Hence I argue: The Third Way is taken from possibility and necessity, and runs thus. We find in nature various regularities, and may regard them as coincidence or as necessarily proceeding from underlying reasons. If the former, we cannot *know* them as regularities, for what happens by coincidence may not happen. But some regularities we know as regularities, hence they proceed from underlying reasons.

And I support this, by reference to two eminent exponents of scientific realism, as well as by appeal to common knowledge. In a famous lecture, C. S. Peirce addressed his audience as follows:

Suppose we attack the question experimentally. Here is a stone. Now I place that stone where there will be no obstacle between it and the floor, and I will predict with confidence that as soon as I let go my hold upon the stone, it will fall to the floor. I will prove that I can make a correct prediction by actual trial if you like. But I see by your faces that you all think it will be a very silly experiment.[9]

And J. J. C. Smart seems to be elaborating this very point when he writes:

If the phenomenalist [i.e. unbeliever] about theoretical entities is correct, we must believe in a *cosmic coincidence*. That is, if this is so, statements about electrons, etc., are of only instrumental value: they simply enable us to predict phenomena on the level of galvanometers and cloud chambers. They would do nothing to remove the *surprising character* of these phenomena. On the other hand, if we interpret a scientific theory in a realist way, then we have no need for such a cosmic coincidence: it is not surprising that galvanometers and cloud chambers behave in the sort of way they do, for if there really are electrons, etc., this is just what we should expect.[10]

While we have learned not to ask the question *why is there a world rather than nothing?*, we still regard as totally legitimate the question *why is the world the way it is rather than some other way?* And the answer *By coincidence*, or *As a matter of fact*, is as unacceptable to the second question as to the first.

IV

The Fourth Way is undoubtedly the most difficult to understand, the most subtle and, just possibly, the most confused; but from a metaphysical point of view, it is yet also the most profound.

Thus Aquinas: The Fourth Way is taken from the gradation to be found in things. Among beings, there are some more and some less good, true, noble, and the like. But *more* and *less* are predicated of things in accordance with their resemblance to something which is the maximum (for example, one thing is hotter than another if the former more nearly resembles fire, which is hottest). So there must be something truest, best, noblest, and consequently something which is greatest in being; for as Aristotle says, what is greatest in truth has most reality.[11]

I must admit that at first, I quite despaired of adapting this argument to the cause of scientific realism. But there is a very important and exceedingly subtle analogous point.[12] To prepare the ground, let me yet again quote Professor Smart:

Indeed, I would wish to go further than merely defend the physicist's picture of the world as an ontologically respectable one. I would wish to urge that the physicist's language gives us a *truer* picture of the world than does the language of ordinary common sense.... Up to a point, Susan Stebbing's strictures against Eddington were perfectly justified. But Stebbing went too far. There is also a perfectly good sense in which it is true and illuminating to say that the table is *not* solid. The atoms which compose the table are like the solar system in being mostly empty space. (This was Eddington's point.) So, though most commonsense propositions in ordinary life are true, I still wish to say that science gives us a 'truer picture' of the world.[13]

And two pages later:

What is needed is not that the micro-theory should explain a macro-theory or macro-laws to which it is linked by correspondence rules. What is needed is, as Wilfrid Sellars and Feyerabend have pointed out, that it should explain why observable things obey, to the extent that they do, those macro-laws.[14]

I must apologize to Professor Smart for quoting disjoint passages from an eminently readable chapter; but I hope that these passages will prepare the reader for my own short exposition of an exceedingly subtle manœuvre.

So far, I have been talking about theoretical entities postulated by scientific theories to explain the regularities in natural phenomena. But you see, strictly speaking, there are no such regularities! I am not here becoming devil's advocate and taking the part of the nominalist or anti-realist, who sees any and all regularities and irregularities as so much happenstance. Even from the point of view of the most hard-necked scientific realist, it would be too much of a coincidence if we could discern more than approximate regularities in the natural phenomena (given that there is a relatively low and very finite limit to the degree of complexity of humanly discernible regularities). Just consider the contents of your pockets or handbag: these contents are under your voluntary control, and you have your reasons for whatever you put in or take out. Yet are there any discernible *strict* regularities in what pockets and handbags contain? Most likely not—*just because* the regularities are on a more basic level.

So what we must explain is not putative regularities in the natural phenomena, but rather, why the phenomena approximate the apparent regularities to the extent they do. And when science describes an underlying structure of greater unity, coherence, simplicity, and regularity than the phenomena could ever hope to have, then this very degree of unity argues that the scientific picture is a truer one, and that the scientific world—and Eddington's table—has greater reality than the world of common sense.

I shall not try to codify this in a formal argument. However, I should like to point out (with reference to the pockets and handbags example), how very secure the realist's position is, after consideration of the Third and Fourth Ways. For if there are regularities in natural phenomena, to whatever extent or degree of exactness, these require the postulation of micro-structure to explain them. On the other hand, if there are no regularities, to any great extent or great degree of exactness, this too shows the truth of realism. For then it is not surprising that galvanometers and cloud chambers do not exhibit any exact regularities: if the basic laws govern electrons, etc., this is just what we would expect.

V

If the Fourth Way is the most profound, the Fifth Way is the most fun. Known popularly as the Argument from Design, it was immortalized in Paley's watch and Pangloss's spectacles.

Thus Aquinas: We see that things which lack knowledge, such as natural bodies, act for an end, and this is evident from their acting always, or nearly always, in the same way, so as to obtain the best result. Hence it is plain that they achieve their end, not fortuitously, but by design. Whatever lacks intelligence cannot move towards an end, unless it be directed by a being endowed with intelligence; as the arrow is directed by the archer. Therefore some intelligent being exists by whom all natural things are directed to their end; and this being we call God.

The strong support of this argument by various eighteenth-century apologists has been richly documented by Hick.[15] I may here refer to Derham's *Physico-Theology* (1713), which notes that nature is so governed that overpopulation problems automatically do not arise, and his *Astro-Theology* (1714), which argues that there must surely be an underlying reason why all planets are round, rather than 'one of this, another of a quite different figure: one square, one multangular, another long, and another of another shape. ...'

Now I maintain that what is wrong with Aquinas's Fifth Way, is *not* his pattern of argument, but rather his premises. For, upon seeing the amount of regularity and structure in the natural phenomena, Aquinas argues that we should *opt for the best explanation.* He adds that the best explanation is that which postulates intelligent and purposive design. This premiss we deny. But that scientific inference is acceptance as true of the best (available) explanation, is a position held, in various forms, by many modern philosophers.

Hence I argue: We see that many things, such as natural bodies, exhibit great regularity in their behaviour, reactions, and evolution. This can be asserted to be so as a matter of happenstance, or explained through the postulation of micro-structure underlying the phenomena. Since it is correct scientific practice to infer to the best explanation, we must take the latter course. And so we must accept, as a literally true representation, the picture disclosed by our best available scientific theories.

Concluding Scientific Postscript

It may have occurred to you, gentle reader—however much you fied the thought—that the new Five Ways may be attacked in analogy to Hume's attack upon the old. This disturbing thought I mean to dispel.

I may sum up Hume's, and others', counterarguments as follows. Granted that the regress in causation, or explanation, must have a terminus, there is no reason why that should not be the universe itself. There is no reason to regard God as a more fitting terminus than the world. For if the world becomes intelligible only through reference to God's will, how shall we understand God's will? And if we cannot understand God's will, why not stop with the universe, which we could not understand in the first place?

All attempts to counter this counterargument seem to consist, essentially, in the assertion that God is fundamentally different from the world. With respect to God, the question of cause or explanation or ground no longer arises. As Hick put it, 'the idea of God provides a *de jure* as well as a *de facto* terminus to the explanatory process.'[16]

However this may be in the case of God, we can see a possible pattern of counterargument for our opposition. He may argue: with respect to explanation, there is no difference between galvanometers and electrons. To postulate a micro-structure exhibiting underlying regularities, is only to posit a new cosmic coincidence. That galvanometers and cloud chambers behave as they do, is still surprising if there are electrons, etc., for it is surprising that there should be such regularity in the behaviour of electrons, etc. If not metaphysically inclined, he will then be happy with the prior coincidence that human appeal to quantum theory brought order to the chaos of galvanometer and cloud-chamber data. For he did not understand that prior coincidence in the first place, and that is enough. On the other hand, if metaphysically inclined, he will ask (about the micro-entities as well): what makes entities of the same constitution behave in the same way, in times past, present, and to come? And a terrible new realist beauty is born.

Against this, I counter that it is regularities in the phenomena only for which the question of explanation arises. If the question seems to arise, why certain theoretical entities behave as they do, this is really a question of a different order. For in that case, there are two possibilities. Either a further and as yet unexplained regularity in the phenomena is discussed, and the theory requires that it be traced back to *those* theoretical entities; or else, it is being conjectured that the theory can be made simpler and more cohesive by an amendment of its postulates. In the first case, the motive is provided by the natural phenomena, and in the second by pragmatic demands for economy of thought. In neither case is it the regularities behind the phenomena which *ipso facto* demand explanation.

Although this counters the objection and brings the exposition to an end, I cannot avoid a short paragraph to correct the popular misconception that the distinction between observable entities and others cannot be drawn—which would fatally undermine the above defence. They ill serve the cause of scientific realism, who argue that no such distinction can be made! Their arguments are of three sorts.[17]

Obj. 1. The distinction between observation through instruments and inference from data cannot be drawn. Can we observe through an electron microscope? Through an optical microscope? Through a reading glass? Through window panes?

Which objection I counter by reducing to absurdity the idea that a difference of degree is no difference. For on that account, everyone is poor: if a man has one penny, he is poor; and if a poor man be given a penny, he is still poor. Hence, by mathematical induction, everyone is poor. I take no credit for this *sorites* sophism, and hesitate to give credit for the Objection it refutes.

Obj. 2. According to modern valency theory, some large crystals are single molecules. But molecules are theoretical entities. Hence some theoretical entities are observable.

The crystal-molecule example is no better for being striking. Congeries of theoretical entities are also theoretical entities; hence if this table is a congeries of subatomic particles, and this table is observable, then some theoretical entities are observable. But *that* is a logically valid argument and therefore trivial. In its insinuations it is not trivial—for these compare it with the analogue: 'Planetarians comprise both Earthmen and Venusians. There goes one now! (pointing to Professor Maxwell) Now do you believe they are real?'

Obj. 3. Theoretical entities and processes tend to become observable, and join the phenomena, soon after their original postulation; for example, germs and viruses.

Which objection I counter by noting the equally regrettable tendency of theoretical entities to go out of existence altogether soon after they are observed. Whose electron did Millikan observe; Lorentz's, Rutherford's, Bohr's, or Schrödinger's? A good example also are the *homunculi*: when van Leeuwenhoek examined his semen under the new microscope, he saw these postulated fully formed little humans swimming around. Not only that, his friends (all male) saw them too.[18]

Since, therefore, the assertion of an important difference (in relevant respects) between theoretical and non-theoretical entities is crucial to foiling the anti-realist counterargument with which I began; and since the objections to the contrary are found wanting; I conclude that there is such a difference and that the new Five Ways cannot be countered like the old.

Notes

CHAPTER 1

[1] See my 'A Re-examination of Aristotle's Philosophy of Science' (*Dialogue*, 1980) and 'Essence and Existence', pp. 1–25, in N. Rescher (ed.), *Studies in Ontology*, American Philosophical Quarterly, Monograph No. 12 (Oxford: Blackwell, 1978), for discussion of some philosophical issues relating to that tradition.

[2] *The Works of the Honourable Robert Boyle*, ed. Birch (London, 1672), vol. III, p. 13; I take the quotation from R. S. Woolhouse, *Locke's Philosophy of Science and of Language* (Oxford: Blackwell, 1971), which has an excellent discussion of the philosophical issues of that period and of Boyle's role.

[3] I. Levi, 'Confirmational Conditionalization', *Journal of Philosophy*, 75 (1978), 730–7; p. 737.

[4] A. Fine, 'How to Count Frequencies: A Primer for Quantum Realists', *Synthese*, 42 (1979); quotation from pp. 151–2.

CHAPTER 2

[1] Brian Ellis, *Rational Belief Systems* (Oxford: Blackwell, 1979), p. 28.

[2] Hartry Field has suggested that 'acceptance of a scientific theory involves the belief that it is true' be replaced by 'any reason to think that any part of a theory is not, or might not be, true, is reason not to accept it.' The drawback of this alternative is that it leaves open what epistemic attitude acceptance of a theory does involve. This question must also be answered, and as long as we are talking about full acceptance—as opposed to tentative or partial or otherwise qualified acceptance—I cannot see how a realist could do other than equate that attitude with full belief. (That theories believed to be false are used for practical problems, for example, classical mechanics for orbiting satellites, is of course a commonplace.) For if the aim is truth, and acceptance requires belief that the aim is served ... I should also mention the statement of realism at the beginning of Richard Boyd, 'Realism, Underdetermination, and a Causal Theory of Evidence', *Noûs*, 7 (1973), 1–12. Except for some doubts about his use of the terms 'explanation' and 'causal relation' I intend my statement of realism to be entirely in accordance with his. Finally, see C. A. Hooker, 'Systematic Realism', *Synthese*, 26 (1974), 409–97; esp. pp. 409 and 426.

[3] More typical of realism, it seems to me, is the sort of epistemology found in Clark Glymour's forthcoming book, *Theory and Evidence* (Princeton: Princeton University Press, 1980), except of course that there it is fully and carefully developed in one specific fashion. (See esp. his chapter 'Why I am not a Bayesian' for the present issue.) But I see no reason why a realist, as such, could not be a Bayesian of the type of Richard Jeffrey, even if the Bayesian position has in the past been linked with antirealist and even instrumentalist views in philosophy of science.

[4] G. Maxwell, 'The Ontological Status of Theoretical Entities', *Minnesota Studies in Philosophy of Science*, III (1962), p. 7.

[5] There is a great deal of recent work on the logic of vague predicates; especially important, to my mind, is that of Kit Fine ('Vagueness, Truth, and Logic', *Synthese*, 30 (1975), 265–300) and Hans Kamp. The latter is currently working on a new theory of vagueness that does justice to the 'vagueness of vagueness' and the context-dependence of standards of applicability for predicates.

[6] Op. cit., p. 15. In the next chapter I shall discuss further how observability should be understood. At this point, however, I may be suspected of relying on modal distinctions which I criticize elsewhere. After all, I am making a distinction between human limitations, and accidental factors. A certain apple was dropped into the sea in a bag of refuse, which sank; relative to that information it is necessary that no one ever observed the apple's core. That information, however, concerns an accident of history, and so it is not human limitations that rule out observation of the apple core. But unless I assert that some facts about humans are essential, or physically necessary, and others accidental, how can I make sense of this distinction? This question raises the difficulty of a philosophical retrenchment for modal language. This I believe to be possible through an ascent to pragmatics. In the present case, the answer would be, to speak very roughly, that the scientific theories we accept are a determining factor for the set of features of the human organism counted among the limitations to which we refer in using the term 'observable'. The issue of modality will occur explicitly again in the chapter on probability.

[7] *Science, Perception and Reality* (New York: Humanities Press, 1962); cf. the footnote on p. 97. See also my review of his *Studies in Philosophy and its History*, in *Annals of Science*, January 1977.

[8] Cf. P. Thagard, doctoral dissertation, University of Toronto, 1977, and 'The Best Explanation: Criteria for Theory Choice', *Journal of Philosophy*, 75 (1978), 76–92.

[9] 'The Inference to the Best Explanation', *Philosophical Review*, 74 (1965), 88–95 and 'Knowledge, Inference, and Explanation', *American Philosophical Quarterly*, 5 (1968), 164–73. Harman's views were further developed in subsequent publications (*Noûs*, 1967; *Journal of Philosophy*, 1968; in M. Swain (ed.), *Induction*, 1970; in H.-N. Castañeda (ed.), *Action, Thought, and Reality*, 1975; and in his book *Thought*, Ch. 10). I shall not consider these further developments here.

[10] See esp. 'Knowledge, Inference, and Explanation', p. 169.

[11] J. J. C. Smart, *Between Science and Philosophy* (New York: Random House, 1968), p. 151.

[12] Ibid., pp. 150f.

[13] This point is clearly made by Aristotle, *Physics*, II, Chs. 4–6 (see esp. 196^a 1–20; 196^b 20–197^a 12).

[14] W. Salmon, 'Theoretical Explanation', pp. 118–45 in S. Körner (ed.), *Explanation* (Oxford: Blackwell, 1975). In a later paper, 'Why ask why?' (Presidential Address, *Proc. American Philosophical Association* 51 (1978), 683–705), Salmon develops an argument for realism like that of Smart's about coincidences, and adds that the demand for a common cause to explain apparent coincidences formulates the basic principle behind this argument. However, he has weakened the common cause principle so as to escape the objections I bring in this section. It seems to me that his argument for realism is also correspondingly weaker. As long as there is no universal demand for a common cause for every pervasive regularity or correlation, there is no *argument* for realism here. There is only an explanation of why it is satisfying to the mind to postulate explanatory, if unobservable, mechanisms *when we can*. There is no argument in that the premises do not *compel* the realist conclusion. Salmon has suggested in conversation that we should perhaps impose the universal demand that only correlations among spatio-temporally (approximately) coincident events are allowed to remain without explanation. I do not see a rationale for this; but also, it is a demand not met by quantum mechanics in which there are non-local correlations (as in the Einstein–Podolski–Rosen 'paradox'); orthodox physics refuses to see these correlations as genuinely paradoxical. I shall discuss Salmon's more recent theory in Ch. 4. These are skirmishes; on a more basic level I wish to maintain that there is sufficient satisfaction for the mind if we can construct theories in whose models correlations and apparent coincidences are traceable back to common

causes—without adding that all features of these models correspond to elements of reality. See further my 'Rational Belief and the Common Cause Principle', in R. McLaughlin's forthcoming collection of essays on Salmon's philosophy of science.

¹⁵ H. Reichenbach, *Modern Philosophy of Science* (London: Routledge and Kegan Paul, 1959), Chs. 3 and 5. From a purely logical point of view this is not so. Suppose we define the predicate $P(-m)$ to apply to a thing at time t exactly if the predicate P applies to it at time $t+m$. In that case, description of its 'properties' at time t, using predicate $P(-m)$, will certainly give the information whether the thing is P at time $t+m$. But such a defined predicate 'has no physical significance', its application cannot be determined by any observations made at or prior to time t. Thus Reichenbach was assuming certain criteria of adequacy on what counts as a description for empirical science; and surely he was right in this.

¹⁶ H. Reichenbach, *The Direction of Time* (Berkeley: University of California, 1963), Sect. 19, pp. 157–63; see also Sects. 22 and 23.

¹⁷ The paper by Einstein, Podolski, and Rosen appeared in the *Physical Review*, 47 (1935), 777–80; their thought experiment and Compton scattering are discussed in Part 1 of my 'The Einstein–Podolski–Rosen Paradox', *Synthese*, 29 (1974), 291–309. An elegant general result concerning the extent to which the statistical 'explanation' of a correlation by means of a third variable requires determinism is the basic lemma in P. Suppes and M. Zanotti, 'On the Determinism of Hidden Variable Theories with Strict Correlation and Conditional Statistical Independence of Observables', pp. 445–55 in P. Suppes (ed.), *Logic and Probability in Quantum Mechanics* (Dordrecht: Reidel Pub. Co., 1976). This book also contains a reprint of the preceding paper.

¹⁸ There is another way: if the correlation between A and B is known, but only within inexact limits, the postulation of the common cause C by a theory which specifies $P(A/C)$ and $P(B/C)$ will then entail an exact statistical relationship between A and B, which can be subjected to further experiment.

¹⁹ See my 'Wilfrid Sellars on Scientific Realism', *Dialogue*, 14 (1975), 606–16; W. Sellars, 'Is Scientific Realism Tenable?', pp. 307–34 in F. Suppe and P. Asquith (eds.), *PSA 1976* (East Lansing, Mich.: Philosophy of Science Association, 1977), vol. II; and my 'On the Radical Incompleteness of the Manifest Image', ibid., 335–43; and see n. 7 above.

²⁰ W. Sellars, 'The Language of Theories', in his *Science, Perception, and Reality* (London: Routledge and Kegan Paul, 1963).

²¹ Op. cit., p. 121.

²² Ibid., p. 121.

²³ Ibid., p. 123.

²⁴ See my 'Semantic Analysis of Quantum Logic', in C. A. Hooker (ed.), *Contemporary Research in the Foundations and Philosophy of Quantum Theory* (Dordrecht: Reidel, 1973), Part III, Sects. 5 and 6.

²⁵ Hilary Putnam, *Philosophy of Logic* (New York: Harper and Row, 1971)—see also my review of this in *Canadian Journal of Philosophy*, 4 (1975), 731–43. Since Putnam's metaphysical views have changed drastically during the last few years, my remarks apply only to his views as they then appeared in his writings.

²⁶ Op. cit., p. 63.

²⁷ Ibid., p. 67.

²⁸ Ibid., p. 69.

²⁹ Hilary Putnam, *Mathematics, Matter and Method* (Cambridge: Cambridge University Press, 1975), vol. I, pp. 69f.

³⁰ Michael Dummett, *Truth and Other Enigmas* (Cambridge, Mass.: Harvard University Press, 1978), p. 146 (see also pp. 358–61).

³¹ Dummett adds to the cited passage that he realizes that his characterization does

not include all the disputes he had mentioned, and specifically excepts nominalism about abstract entities. However, he includes scientific realism as an example (op. cit., pp. 146f.).

[32] This is especially relevant here because the 'translation' that connects Putnam's two foundations of mathematics (existential and modal) as discussed in this essay, is not a literal construal: it is a mapping presumably preserving statementhood and theoremhood, but it does not preserve logical form.

[33] Putnam, op. cit., p. 73 (n. 29 above). The argument is reportedly developed at greater length in Boyd's forthcoming book *Realism and Scientific Epistemology* (Cambridge University Press).

[34] Of course, we can ask specifically why the *mouse* is one of the surviving species, how *it* survives, and answer this, on the basis of whatever scientific theory we accept, in terms of its brain and environment. The analogous question for theories would be why, say, Balmer's formula for the line spectrum of hydrogen survives as a successful hypothesis. In that case too we explain, on the basis of the physics we accept now, why the spacing of those lines satisfies the formula. Both the question and the answer are very different from the global question of the success of science, and the global answer of realism. The realist may now make the *further* objection that the anti-realist cannot answer the question about the mouse specifically, nor the one about Balmer's formula, in this fashion, since the answer is in part an assertion that the scientific theory, used as basis of the explanation, is true. This is a quite different argument, which I shall take up in Ch. 4, Sect. 4, and Ch. 5.

In his most recent publications and lectures Hilary Putnam has drawn a distinction between two doctrines, metaphysical realism and internal realism. He denies the former, and identifies his preceding scientific realism as the latter. While I have at present no commitment to either side of the metaphysical dispute, I am very much in sympathy with the critique of Platonism in philosophy of mathematics which forms part of Putnam's arguments. Our disagreement about scientific (internal) realism would remain of course, whenever we came down to earth after deciding to agree or disagree about metaphysical realism, or even about whether this distinction makes sense at all.

CHAPTER 3

[1] This chapter is partly based on my paper by the same name in the *Journal of Philosophy*, 73 (1976), 623–32, presented to the American Philosophical Association, Boston, December 1976, with commentary by Richard Boyd and Clark Glymour.

[2] A1–A3 are essentially Hilbert's 'axioms of connection' for points and lines: see D. Hilbert, *The Foundations of Geometry*, trans. E. J. Townsend (Chicago: Open Court Publishing Co., 1902), Ch. I, Sect. 1. For an original and comprehensive discussion of models both in mathematics and the sciences see P. Suppes, 'A Comparison of the Meaning and Uses of Models in Mathematics and the Empirical Sciences', *Synthese*, 12 (1960), 287–301.

[3] F. Cajori (ed.), *Sir Isaac Newton's Mathematical Principles of Natural Philosophy and His System of the World* (Berkeley: University of California Press, 1960), p. 12.

[4] Herbert A. Simon, 'The Axiomatization of Classical Mechanics', *Philosophy of Science*, 21 (1954), 340–3.

[5] Op. cit., Book III *Of the System of the World*, Hypothesis I, Prop. XI, and Cor. Prop. XII.

[6] Cf. Richard N. Boyd, 'Realism, Underdetermination and a Causal Theory of Evidence', *Noûs*, 7 (1973), 1–12.

[7] Taken from James Clerk Maxwell, 'A Dynamical Theory of the Electromagnetic Field', *Philosophical Transactions*, 155 (1865); the passage is found on p. 293 in the

partial reprint of this paper in M. H. Shamos, *Great Experiments in Physics* (New York: Holt and Co., 1959).

[8] Henri Poincaré, *The Value of Science*, trans. B. Halsted (New York: Dover, 1958), p. 98. As Clark Glymour has pointed out, in unpublished notes, a line of reasoning of the sort examined here occurs in J. Earman and M. Friedman, 'The Meaning and Status of Newton's Law of Inertia and the Nature of Gravitational Forces', *Philosophy of Science*, 40 (1973), 329–59.

[9] *Historical Development of Logic*, trans. J. Rosenthal (New York: Henry Holt and Co., 1929), p. 230.

[10] Brian Ellis, 'The Origins and Nature of Newton's Laws of Motion', pp. 29–68, in R. Colodny (ed.), *Beyond the Edge of Certainty* (Englewood Cliffs, N.J.: Prentice-Hall, 1965).

[11] F. J. Belinfante, *A Survey of Hidden-Variable Thories* (New York: Pergamon Press, 1973), pp. 25f.

[12] I state Craig's result in the form in which it was mainly utilized in the philosophical discussions; the result is more general and need not rely on a division of vocabulary. The literature on this topic is large but it suffices to consult the exposition of its fallacies by C. A. Hooker, 'Craigian Transcriptionism', *American Philosophical Quarterly*, 5 (1968), 152–63, and 'Five Arguments Against Craigian Transcriptionism', *Australasian Journal of Philosophy*, 46 (1968), 265–76.

[13] For example, David Lewis, 'How to Define Theoretical Terms', *Journal of Philosophy*, 67 (1970), 427–46. This paper is an example of the use of the syntactic scheme, but is not subject to my other criticisms. On the contrary, *correctly read*, it provides independent reasons for the conclusion that the empirical import of a theory cannot be syntactically isolated.

[14] See n. 6 above. Not surprisingly, this paper too has evidence that the empirical import of a theory cannot be syntactically isolated. But Boyd concludes more than that the new terms are as well understood as the old, namely, that there is no distinction between truth and empirical adequacy for scientific theories. See also M. Gardner, 'The Unintelligibility of "Observational Equivalence"', pp. 104–16, in F. Suppe and P. Asquith (eds.), *PSA 1976* (East Lansing, Mich.: Philosophy of Science Association, 1976), Vol. 1. On the other hand, C. A. Hooker (op. cit., pp. 415, 445f., 485) who is a realist, takes the same position on observability as I do; we presented this view independently in papers at the Canadian Philosophical Association in 1974.

[15] See Chs. 8 and 9 of Max Jammer, *Concepts of Mass* (Cambridge, Mass.: Harvard University Press, 1961). Of special interest is the work of Pendse (reported pp. 98–100) which was concerned to delimit the exact extent to which mass is determinable from other quantities.

[16] *Introduction to Logic* (Princeton: Van Nostrand, 1957), p. 298. See further the discussion in the symposium by Bressan, Suppes, and myself on modal concepts in science, in Schaffner, K. F., and R. S. Cohen (eds.), *PSA 1972* (Dordrecht: Reidel, 1974), pp. 285–330.

[17] For all but Mackey's, see Jammer, op. cit., Ch. 9; and G. W. Mackey, *The Mathematical Foundations of Quantum Mechanics* (New York: Benjamin, 1973), pp. 1–4.

[18] Y. Aharonov and L. Susskind, *Phys. Rev.*, 158 (1967), 1237; H. J. Bernstein, *Phys. Rev. Letters*, 18 (1967), 1102; G. C. Hegefeldt and K. Krauss, *Phys. Rev.*, 170 (1968), 1185; R. Mirman, *Phys. Rev.*, D1 (1970), 3349; A. G. Klein and G. I. Opat, *Phys. Rev.*, D11 (1975), 523–8 and *Phys. Rev. Letters*, 37 (1976), 238–40. I have benefited here from discussion with Prof. E. Levy, University of British Columbia and Dean J. Marburger, University of Southern California.

[19] N. Cartwright, 'Superposition and Macroscopic Observation', pp. 231–44, in P. Suppes (ed.), *Logic and Probability in Quantum Mechanics* (Dordrecht: Reidel, 1976). The article appeared previously in *Synthese*, 29 (1974), 229–42. See also Ch. 1

of F. J. Belinfante, *Measurement and Time Reversal in Objective Quantum Mechanics* (New York: Pergamon Press, 1976).

[20] C. Glymour, 'Cosmology, Convention, and the Closed Universe', *Synthese*, 24 (1972), 195–218; discussed in my 'Earman on the Causal Theory of Time', ibid., 87–95. (This published paper by Glymour is essentially the same as the unpublished one to which I referred.)

[21] C. Glymour, 'Indistinguishable Space–Times and the Fundamental Group', in J. Earman, C. Glymour, and J. Stachel (eds.), *Minnesota Studies in the Philosophy of Science*, 8 (Minneapolis: University of Minnesota, 1977).

[22] See for example, R. Wojcicki, 'Set Theoretic Representations of Empirical Phenomena', *Journal of Philosophical Logic*, 3 (1974), 337–43; M. Przełeski, *The Logic of Empirical Theories* (London: Routledge and Kegan Paul, 1969); M. L. Dalla Chiara and G. Toraldo di Francia, 'A Logical Analysis of Physical Theories', *Rivista di Nuovo Cimento*, Serie 2, 3 (1973), 1–20—a much more extensive presentation is to be found in the proceedings of the Enrico Fermi Institute Summer School on Foundations of Physics held in Varenna, 1977; (see Ch. 6, n. 16 below); P. Suppes, 'Models of Data' and 'Measurement, Empirical Meaningfulness, and Three-Valued Logic', in his *Studies in Methodology and Foundations of Science* (Dordrecht: Reidel, 1969), and 'What is a Scientific Theory', pp. 55–67 in S. Morgenbesser (ed.), *Philosophy of Science Today* (New York: Basic Books, 1967); F. Suppe (ed.), *The Structure of Scientific Theories* (Urbana: University of Illinois Press, 1974), Introduction, pp. 221–30, and 'Theories, their Formulations, and the Operational Imperative', *Synthese*, 25 (1973), 129–64. (For further discussion, see the exchange between Przełeski and Tuomela in *Synthese*, 25 (1972) and 26 (1974).) I am claiming agreement with these authors only with respect to the role of empirical substructures in models.

[23] See Stanley Gudder, 'Hidden Variables in Quantum Mechanics Reconsidered', *Review of Modern Physics*, 40 (1968), 229–31; and Sect. III of my 'Semantic Analysis of Quantum Logic', pp. 80–113 in C. A. Hooker (ed.), *Contemporary Research in the Foundations and Philosophy of Quantum Theory* (Dordrecht: Reidel, 1973); and Belinfante, op. cit. (n. 11 above).

CHAPTER 4

[1] I gave a preliminary exposition of an epistemological position compatible with nominalism and empiricism as I understand them in my 'Rational Belief and Belief Change: the Dynamics of Faith', presented at the Canadian Philosophical Association, London, Ontario, June 1978.

[2] The view advanced here is clearly at odds with the one generally called Bayesian and associated with de Finetti's subjectivist interpretation of philosophy, as espoused, for example, by Richard Jeffrey. On the question of acceptance of hypotheses as a rational procedure actually instantiated in science, I agree largely with such realists as Clark Glymour and Ronald Giere. This is not to say that a Bayesian view could not be consistently meshed with constructive empiricism. Note that Glymour similarly does not deny that a realist could use a Bayesian epistemology to make sense of scientific methodology; but he adds, after all, 'both bears and ballerinas can dance' (*Theory and Evidence*, forthcoming).

[3] The role of theory in testing, and hence in experimental design, is of course a subject of some complexity. It is explored most interestingly and comprehensively so far in Clark Glymour's 'bootstrap' account of the relation of evidence to theory, developed in his article 'Relevant Evidence', *Journal of Philosophy*, 62 (1975), 403–26, and his forthcoming book *Theory and Evidence* (see also his exchange with Paul Horwich in the *Journal of Philosophy*, 63 (1978)). While Glymour develops his account from a realist point of view, and indeed holds that the evidence we have may better support one theory than another, although as a matter of fact both are empirically

adequate, I have as yet seen no reason to think that his account could not be adapted to anti-realist views. Specifically, if one theory is better supported by the evidence than another, in his sense, we may have greater reason to believe that it is empirically adequate or that it can be extended to an empirically adequate theory for a larger domain.

⁴ 'Realism, Underdetermination, and a Causal Theory of Evidence', *Nous*, 7 (1973), 1–12.

⁵ In 'Explanation and Reference', pp. 199–221, in G. Pearce and P. Maynard (eds.), *Conceptual Change* (Dordrecht: Reidel, 1973), and reprinted as Ch. II of his *Mind, Language, and Reality: Philosophical Papers*, vol. II (Cambridge: Cambridge University Press, 1975), Hilary Putnam first offered the conjunction objection as an argument that there was no positivist replacement for the concept of truth. In 'Reference and Understanding', Part 3 of his *Meaning and the Moral Sciences* (London: Routledge and Kegan Paul, 1978), however, he refers to it as an argument to the effect that 'an account of this kind—an account that says that what we seek is a kind of acceptability that lacks the property of deductive closure—fails to justify the norms of scientific practice' (p. 102). The objection was emphasized to me in conversation by Richard Boyd, Clark Glymour, and Christopher Peacocke.

⁶ I take the term 'mini-theory' from mimeographed notes circulated in 1977 by Edwin Levy (University of British Columbia) who has convinced me of the need for philosophy of science to study this aspect of scientific theorizing in its own right.

⁷ I learned this way of putting it from Clark Glymour's commentary on my paper 'The Pragmatics of Explanation' at the American Philosophical Association (Pacific Division), Portland, March 1977.

⁸ 'Foundations of the Theory of Signs', in O. Neurath, R. Carnap, and C. Morris (eds.), *Foundations of the Unity of Science: Towards an International Encyclopedia of Unified Science* (Chicago: University of Chicago Press, 1955), vol. I, pp. 73–137.

⁹ *The Structure of Science* (New York: Harcourt, Brace, and World, 1961), p. 4; see also ibid., pp. viii, 5, 15.

¹⁰ 'Realism and Instrumentalism', pp. 280–308, in M. Bunge (ed.), *The Critical Approach to Science and Philosophy* (New York: Free Press, 1964).

¹¹ General Scholium to Book III, *Mathematical Principles of Natural Philosophy*, trans. A. Motte (London: Dawsons of Pall Mall, 1968), vol. II, p. 392.

¹² Cf. M. Fierz, 'Does a physical theory comprehend an "objective, real, single process"?', pp. 93–6 in S. Körner (ed.), *Observation and Interpretation* (New York: Academic Press, 1957).

¹³ See Belinfante, op. cit. (Ch. 3, n. 11).

CHAPTER 5

¹ This chapter is based in part on my paper by the same title, *American Philosophical Quarterly*, 14 (1977), 143–50, presented to the American Philosophical Association, Portland, March 1977, with commentary by Kit Fine and Clark Glymour.

² A. Lavoisier, *Œuvres* (Paris: Imp. Imperiale, 1862), vol. II, p. 640. I owe this and the other historical references below to my former student, Paul Thagard.

³ C. Darwin, *On the Origin of the Species* (text of 6th edn., New York: Colllier, 1962), p. 476.

⁴ Cf. Christiaan Huygens, *Treatise on Light*, trans. S. P. Thompson (New York: Dover, 1962), pp. 19f., 22, 63; Thomas Young, *Miscellaneous Works*, ed. G. Peacock (London: John Murray, 1855), vol. I, pp. 168, 170.

⁵ A. Fresnel, *Œuvres complètes* (Paris: Imp. Imperiale, 1866), vol. I, p. 36 (see also pp. 254, 355); Lavoisier, op. cit., p. 233.

⁶ C. Darwin, *The Variations of Animals and Plants* (London: John Murray, 1868),

vol. I, p. 9; *On the Origin of the Species* (Facs. of the first edition, Cambridge, Mass.: Harvard, 1964), p. 408.

[7] C. G. Hempel and P. Oppenheim, 'Studies in the Logic of Explanation', *Philosophy of Science*, 15 (1948), 135–75.

[8] C. G. Hempel, *Philosophy of Natural Science* (Englewood Cliffs, N.J.: 1966), pp. 48f.; see S. Bromberger, 'Why-Questions', (n. 32 below) for some of the counter-examples.

[9] W. C. Salmon, *Statistical Explanation and Statistical Relevance* (Pittsburgh: University of Pittsburgh Press, 1971), pp. 33f.

[10] M. Beckner, *The Biological Way of Thought* (Berkeley: University of California Press, 1968), p. 165 (first pub. 1959, by Columbia University Press).

[11] In a presentation at a conference at the University of Illinois; a summary of the paper may be found in F. Suppe (ed.), *The Structure of Scientific Theories* (Urbana, Ill.: University of Illinois Press, 1974).

[12] Salmon, op. cit., p. 64.

[13] Nancy Cartwright, 'Causal Laws and Effective Strategies', *Noûs*. 1979; the examples cited had been communicated in 1976.

[14] Salmon, op. cit., p. 78. For reasons of exposition, I shall postpone discussion of the *screening off* relation to Sect. 2.6, although Salmon already used it here.

[15] M. Friedman, 'Explanation and Scientific Understanding', *Journal of Philosophy*, 71 (1974), 5–19.

[16] See P. Kitcher, 'Explanation, Conjunction, and Unification', *Journal of Philosophy*, 73 (1976), 207–12.

[17] J. Greeno, 'Explanation and Information', pp. 89–104, in W. C. Salmon, op. cit. (n. 9).

[18] T. Kuhn, *The Structure of Scientific Revolutions* (Chicago: University of Chicago Press, 1970), pp. 107f.

[19] P. J. Zwart, *Causaliteit* (Assen: van Gorcum, 1967), p. 133.

[20] J. L. Mackie, 'Causes and Conditions', *American Philosophical Quarterly*, 2 (1965), 245–64. Since then, Mackie has published a much more extensive theory of causation in *The Cement of the Universe* (Oxford: Clarendon Press, 1974). Since I am of necessity restricted to a small and selective, i.e. biased, historical introduction to my own theory of explanation, I must of necessity do less than justice to most authors discussed.

[21] D. Lewis, 'Causation', *Journal of Philosophy*, 70 (1973), 556f.

[22] See N. Goodman, *Fact, Fiction and Forecast* (Cambridge, Mass.: Harvard University Press, 1955), Ch. 1. The logical theory of counterfactual conditionals has been developed, with success in basic respects but in others still subject to debate, in a number of articles of which the first was Robert Stalnaker, 'A Theory of Conditionals', pp. 98–112, in N. Rescher (ed.), *Studies in Logical Theory* (Oxford: Blackwell, 1968), and one book (David Lewis, *Counterfactuals* (Oxford: Blackwell, 1973)). For a summary of the results, and problems, see my 'Report on Conditionals', *Teorema*, 5 (1976), 5–25.

[23] *Mind*, N.S. 3 (1894), 436–8; P. E. B. Jourdain (ed.), *The Philosophy of Mr. B*rtr*nd R*ss*ll* (London: Allen and Unwin, 1918), p. 39. The analysis of the example which I give here is due to Richmond Thomason.

[24] W. C. Salmon, 'Why ask "Why"?', presidential address to the Pacific Division of the American Philosophical Association, San Francisco, March 1978. Page references are to the manuscript version completed and circulated in May 1978; the paper is published in *Proceedings and Addresses of the American Philosophical Association*, 51 (1978), 683–705.

[25] Op. cit., pp. 29f.

[26] Ibid., pp. 14f.

[27] Hans Reichenbach, *The Direction of Time* (Berkeley: University of California Press, 1956), Sects. 19 and 22.

[28] Salmon, op. cit., p. 13.

[29] But it might defeat the use of Salmon's theory in metaphysical arguments, for example, his argument for realism at the end of this paper.

[30] This survey is found in Zwart, op. cit., p. 135, n. 19; references are to Beck's and Nagel's papers in H. Feigl and M. Brodbeck (eds.), *Readings in the Philosophy of Science* (New York: Appleton–Century–Crofts, 1953), pp. 374 and 698, R. B. Braithwaite, *Scientific Explanation* (Cambridge: Cambridge University Press, 1953), p. 320; D. Bohm, *Causality and Chance in Modern Physics* (London: Routledge & Kegan Paul, 1957), *passim*.

[31] N. R. Hanson, *Patterns of Discovery* (Cambridge: Cambridge University Press, 1958), p. 54.

[32] Zwart, op. cit., p. 136; my translation.

[33] S. Bromberger, 'Why-Questions', pp. 86–108, in R. G. Colodny (ed.), *Mind and Cosmos* (Pittsburgh: University of Pittsburgh Press, 1966).

[34] 'Explanations-of-What?', mimeographed and circulated, Stanford University, 1974. The idea was independently developed, by Jon Dorling in a paper circulated in 1976, and reportedly by Alan Garfinkel in *Explanation and Individuals* (Yale University Press, forthcoming). I wish to express my debt to Bengt Hannson for discussion and correspondence in the autumn of 1975 which clarified these issues considerably for me.

[35] For a fuller account of Aristotle's solution of the asymmetries, see my 'A Re-examination of Aristotle's Philosophy of Science', *Dialogue*, 1980. The story was written in reply to searching questions and comments by Professor J. J. C. Smart, and circulated in November 1976.

[36] At the end of my 'The Only Necessity is Verbal Necessity', *Journal of Philosophy*, 74 (1977), 71–85 (itself an application of formal pragmatics to a philosophical problem), there is a short account of the development of these ideas, and references to the literature. The paper 'Demonstratives' by David Kaplan which was mentioned there as forthcoming, was completed and circulated in mimeo'd form in the spring of 1977; it is at present the most important source for the concepts and applications of formal pragmatics, although some aspects of the form in which he develops this theory are still controversial (see also n. 30 to Ch. 6 below).

[37] Belnap's theory was first presented in *An analysis of questions: preliminary report* (Santa Monica, Cal.: System Development Corporation, technical memorandum 7-1287-1000/00, 1963), and is now more accessible in N. D. Belnap Jr. and J. B. Steel, Jr., *The Logic of Questions and Answers* (New Haven: Yale University Press, 1976).

[38] I heard the example from my former student Gerald Charlwood. Ian Hacking and J. J. C. Smart told me that the officer was Sir Charles Napier.

[39] C. L. Hamblin, 'Questions', *Australasian Journal of Philosophy*, 36 (1958), 159–68.

[40] The defining clause is equivalent to 'any proposition which is true if any direct answer to Q is true'. This includes, of course, propositions which would normally be expressed by means of 'metalinguistic' sentences—a distinction which, being language-relative, is unimportant.

[41] In the book by Belnap and Steel (see n. 37 above), Bromberger's theory of why-questions is cast in the general form common to elementary questions. I think that Bromberger arrived at his concept of 'abnormic law' (and the form of answer exhibited by '"Grünbaum" is spelled with an umlaut because it is an English word borrowed from German, and no English words are spelled with an umlaut except those borrowed from another language in which they are so spelled'), because he ignored

the tacit *rather than* (contrast-class) in why-interrogatives, and then had to make up for this deficiency in the account of what the answers are like.

[42] I call this a matter of regimentation, because the theory could clearly be developed differently at this point, by building the claim of relevance into the answer as an explicit conjunct. The result would be an alternative theory of why-questions which, I think, would equally save the phenomena of explanation or why-question asking and answering.

[43] I mention Salmon because he does explicitly discuss this problem, which he calls *the problem of the reference class*. For him this is linked with the (frequency) interpretation of probability. But it is a much more general problem. In deterministic, non-statistical (what Hempel called a deductive–nomological) explanation, the adduced information implies the fact explained. This implication is relative to our background assumptions, or else those assumptions are part of the adduced information. But clearly, our information that the fact to be explained is actually the case, and all its consequences, must carefully be kept out of those background assumptions if the account of explanation is not to be trivialized. *Mutatis mutandis* for statistical explanations given by a Bayesian, as is pointed out by Glymour in his *Theory and Evidence*.

[44] I chose the notation $K(Q)$ deliberately to indicate the connection with models of rational belief, conditionals, and hypothetical reasoning, as discussed for example by William Harper. There is, for example, something called the Ramsey test: to see whether a person with total beliefs K accepts that if A then B, he must check whether $K(A)$ implies B, where $K(A)$ is the 'least revision' of K that implies A. In order to 'open the question' for A, such a person must similarly shift his beliefs from K to $K?A$, the 'least revision' of K that is consistent with A; and we may conjecture that $K(A)$ is the same as $(K?A)\&A$. What I have called $K(Q)$ would, in a similar vein, be a revision of K that is compatible with every member of the contrast class of Q, and also with the denial of the topic of Q. I don't know whether the 'least revision' picture is the right one, but these suggestive similarities may point to important connections; it does seem, surely, that explanation involves hypothetical reasoning. Cf. W. Harper, 'Ramsey Test Conditionals and Iterated Belief Change', pp. 117–35, in W. Harper and C. A. Hooker, *Foundations of Probability Theory, Statistical Inference, and Statistical Theories of Science* (Dordrecht: Reidel, 1976), and his 'Rational Conceptual Change', in F. Suppe and P. Asquith (eds.), *PSA 1976* (East Lansing: Philosophy of Science Association, 1977).

[45] See my 'Facts and Tautological Entailment', *Journal of Philosophy*, 66 (1969), 477–87 and reprinted in A. R. Anderson and N. D. Belnap, Jr., *Entailment* (Princeton: Princeton University Press, 1975), and 'Extension, Intension, and Comprehension', M. Munitz (ed.), *Logic and Ontology* (New York: New York University Press, 1973).

[46] For this and other approaches to the semantics of relevance see Anderson and Belnap, op. cit. (n. 45 above).

[47] I. J. Good, 'Weight of Evidence, Corroboration, Explanatory Power, and the Utility of Experiments', *Journal of the Royal Statistical Society*, series B, 22 (1960), 319–31; and 'A Causal Calculus', *British Journal for the Philosophy of Science*, 11 (1960/61), 305–18 and 12 (1961/62), 43–51. For discussion, see W. Salmon, 'Probabilistic Causality', *Pacific Philosophical Quarterly*, 1980.

CHAPTER 6

[1] Herman Weyl, 'The Ghost of Modality', pp. 278–303 in M. Farber (ed.), *Philosophical Essays in Memory of Edmund Husserl* (Cambridge, Mass.: Harvard University Press, 1940). For a general discussion see my 'Modality', in H. E. Kyburg,

Jr. (ed.), *Current Research in Philosophy of Science* (Lansing, Mich.: Philosophy of Science Association, 1979).

[2] Michael Friedman's forthcoming book on space–time theories defends this view, which has been advocated strongly by John Earman and Clark Glymour (although it seems to me that the latter's theory of evidential support and underdetermination make an anti-realist position possible for him). For an elaboration of the opposite view, and critique of the reification of space–time, see Adolf Grünbaum, 'Absolute and Relational Theories of Space and Space–Time', pp. 303–73, in J. Earman, C. Glymour, and J. Stachel (eds.), *Foundations of Space–Time Theories: Minnesota Studies in the Philosophy of Science*, vol. VIII (Minneapolis: University of Minnesota Press, 1977), and also my *An Introduction to the Philosophy of Time and Space* (New York: Random House, 1970).

[3] H. Everett III, '"Relative State" Formulation of Quantum Mechanics', *Review of Modern Physics*, 29 (1957), 454–62; B. S. De Witt, *The Many Worlds Interpretation of Quantum Mechanics* (Princeton: Princeton University Press, 1973).

[4] David Lewis, *Counterfactuals* (Cambridge, Mass.: Harvard University Press, 1973) and 'How to Define Theoretical Terms' (see n. 13 of Ch. 3 above). Lewis's views on laws of nature are much more sophisticated than my brief remark here indicates, and closer in some ways to views associated with C. S. Peirce and Wilfrid Sellars—see his discussion of Ramsey in *Counterfactuals*, Sect. 3.3. See also the discussion of Lewis's view, and advocacy of an empiricist, non-realist position concerning necessity in Wesley Salmon's new foreword to Hans Reichenbach, *Laws, Modalities and Counterfactuals* (Berkeley: University of California Press, 1976).

[5] Originally proposed by Karl Popper, the propensity interpretation of probability is at present ably defended by a number of philosophers (usually in some combination with a subjective interpretation for some uses of probability—a distinction is made between *objective chance* and *degree of belief*). See especially Hugh Mellor, *The Matter Of Chance* (Cambridge: Cambridge University Press, 1971); Ian Hacking, 'Propensities, Statistics, and Inductive Logic', pp. 485–500, in P. Suppes, *et al.* (eds.), *Logic, Methodology and the Philosophy of Science IV* (Amsterdam: North Holland, 1973); Ronald N. Giere, 'Objective Single Case Probabilities and the Foundations of Statistics', ibid., pp. 467–83, and 'A Laplacean Formal Semantics for Single Case Propensities', *Journal of Philosophical Logic*, 5 (1976), 321–53. For a discussion of such views on probability see Wolfgang Stegmüller, *Personelle und Statistische Wahrscheinlichkeit* (Berlin: Springer-Verlag, 1973) and my review thereof in *Philosophy of Science*, 45 (1978), 158–63.

[6] Henri Poincaré, *Science and Hypothesis*, repr. as part of his *The Foundations of Science* (New York: The Science Press, 1913), Ch. XI, esp. pp. 167f.

[7] For this section I am indebted to the clear analyses given by three authors: Hans Reichenbach, *The Direction of Time* (Berkeley: University of California Press, 1956), Ch. 3; Adolf Grünbaum, *Philosophical Problems of Space and Time*, 2nd enlarged edn. (Dordrecht: Reidel, 1973), Chs. 8, 19; and Henry E. Kyburg, Jr., *The Logical Foundations of Statistical Inference* (Dordrecht: Reidel, 1974), and 'Chance', *Journal of Philosophical Logic*, 5 (1976), 355–93. The term *epistemic probability* is Kyburg's. My adoption of Kyburg's framework for this exposition is not meant to suggest that the relevant distinctions could not be drawn given a Bayesian view of personal probability. In their view, the statistical syllogism is of course only approximately correct.

[8] The interpretation of quantum mechanics which I have advocated is the one I have called the Copenhagen variant of the modal interpretation; it is meant to be a precise version of the so-called orthodox statistical, or Copenhagen, interpretation. However, in these pages I shall try to rely on such features as are accepted by all more or less orthodox interpretations, though not perhaps by those who subscribe either to hidden variables (like Arthur Fine) or the 'quantum-logical' interpretation

associated with Finkelstein, Putnam, Bub, Demopoulos. For exposition of the interpretation I advocate (and arguments that it is a precise, consistent formulation of the main ideas in the 'Copenhagen' or 'orthodox statistical' interpretation) see my 'Semantic Analysis of Quantum Logic' (Ch. 3, n. 23, above), 'The Einstein–Podolski–Rosen Paradox' (Ch. 2, n. 17, above), 'A Semantic Analysis of Niels Bohr's Philosophy of Quantum Theory' (with C. A. Hooker), pp. 221–41, in W. Harper and C. A. Hooker (eds.), *Foundations of Probability Theory, Statistical Inference, and Statistical Theories of Science*, vol. III (Dordrecht: Reidel, 1976). See further the exchange between Fine, Healey, and myself in *Synthese*, 42 (1979), pp. 121–65.

[9] H. Reichenbach, 'The Principle of Anomaly in Quantum Mechanics', *Dialectica*, 2 (1948), 337–50; and see further Nancy Cartwright, 'A Dilemma for the Traditional Interpretation of Quantum Mixtures', pp. 251–8 in K. F. Schaffner and R. S. Cohen (eds.), *PSA 1972* (Dordrecht: Reidel, 1974). Cartwright holds that in certain physical situations, the Ignorance Interpretation of the mixed state is correct (and we have more information than is conveyed by the density matrix) while in other situations it is incorrect. See further Part II of my 'A Formal Approach to Philosophy of Science', pp. 303–66, in R. Colodny (ed.), *Paradigms and Paradoxes* (Pittsburgh: University of Pittsburgh Press, 1972).

[10] In the rejection of the Ignorance Interpretation of Mixtures (and also of the Projection Postulate), I follow Henry Margenau; see his 'Measurement and Quantum States', *Philosophy of Science*, 30 (1963), 1–16 and 138–57, and 'Measurements in Quantum Mechanics', *Annals of Physics*, 23 (1963), 469–85; also his student J. L. Park, 'Quantum Theoretical Concepts of Measurement', *Philosophy of Science*, 35 (1968), 205–31 and 389–41, and 'Nature of Quantum States', *American Journal of Physics*, 36 (1968), 211–26. The issues about mixed states and the alternatives were clearly presented in C. A. Hooker, 'The Nature of Quantum Mechanical Reality: Einstein versus Bohr', pp. 67–302, in R. Colodny (ed.), op. cit. (n. 9 above). See also Neal Grossman, 'The Ignorance Interpretation Defended', *Philosophy of Science*, 41 (1974), 333–44, which is a reply, on this issue, to my papers mentioned above. The defence proposes (as Hooker had suggested, op. cit., pp. 102–5), in effect, that a system which is at present interacting, or has in the past interacted, with another system, be in general not attributed any state at all of its own.

[11] See the Appendix to my 'A Formal Approach . . .' (n. 9 above) and the discussion of Schrödinger's 1935–6 papers, in my 'The Einstein–Podolski–Rosen Paradox' (n. 8 above).

[12] The mixed states include the pure states as a special case. In the general case we use a density matrix or statistical operator to represent states; when that statistical operator is the projection along a single vector ψ, the state is pure. But that pure state is then equivalently represented by the vector itself.

[13] See Max Jammer, *The Philosophy of Quantum Mechanics* (New York: John Wiley and Sons, 1974), pp. 38–44. I rely also on the exposition of Born's work in an unpublished paper by my former student Katherine Arima.

[14] See Part 2 ('The Measurement Problem of Quantum Mechanics as a Consistency Problem') of my 'A Formal Approach . . .' (n. 9 above) and Jeffrey Bub, 'The Measurement Problem of Quantum Mechanics', in the volume mentioned in n. 16 below.

[15] See B. C. van Fraassen and C. A. Hooker, 'A Semantic Analysis of Niels Bohr's Philosophy of Quantum Theory' (see n. 8 above).

[16] This section is based on my 'Relative Frequencies', *Synthese*, 34 (1977), 133–66, which presents an earlier version (now revised) of my proposed interpretation; and more closely on my lectures on the foundations of probability at the Enrico Fermi Institute Summer School on Foundations of Physics (Varenna, 1977) publ. in G. Toraldo di Francia (ed.) *Problems in the Foundations of Physics* (Amsterdam: North-Holland, 1979).

[17] Felix Hausdorff, *Grundzüge der Mengenlehre* (New York: Chelsea Publishing Co., 1949), Ch. X, Sect. 1 (countable addivity), pp. 399–403, and *Nachträge*, pp. 469–72. I owe the references to Gregory H. Moore, and especially thank him for pointing out the result about finitely additive functions in Hausdorff's appendix.

[18] Hans Reichenbach, *The Theory of Probability* (Berkeley: University of California Press, 1949).

[19] This difficulty was pointed out by G. Birkhoff, *Lattice Theory* (Providence: American Mathematical Society, 1940), Ch. XII, Sect. 5 and B. de Finetti, *Probability, Induction, and Statistics* (New York: 1972), Sect. 5.22.

[20] See de Finetti, ibid., Sect. 5.8 and P. Suppes, *Set-Theoretical Structures in Science* (mimeo'd, Stanford, 1967).

[21] R. von Mises, *Mathematical Theory of Probability and Statistics* (New York: 1964), pp. 18–20. The exact statement of Polya's result is this: let $\{p_i\}$ be a set of non-negative real numbers summing to 1, each p_i being the limit of a series of non-negative fractions m_i/m, as m goes to infinity (with m_i a function of m and i). In this supposition, i ranges over the positive integers; let I furthermore be some set of such integers. Then the sum of the numbers p_i with i in I, exists and equals the limit of the sums of m_i/m, with i in I, as m goes to infinity.

[22] Karl Popper, 'The Propensity Interpretation of the Calculus of Probability, and Quantum Mechanics', pp. 65–70, in S. Körner (ed.), *Observation and Interpretation* (New York: Academic Press, 1957); quotation from p. 67.

[23] This use of statistical methods must be classified under the heading of epistemic probability (which covers hypothesis testing, confirmation, and the theory of errors). Every doctrine of objective chance, whether propensity or frequency, must be supplemented by some view of probability as degree of belief—that is, some view concerning the use of probability functions as measures of belief and ignorance.

[24] See the references in n. 5 above.

[25] H. E. Kyburg, Jr., 'Propensities and Probabilities', in *British Journal for the Philosophy of Science*, 25 (1974), 358–75.

[26] See n. 3 above; also n. 8; and the articles by Aldo Bressan, Patrick Suppes, and myself mentioned in n. 16 to Ch. 3.

[27] I wish to thank Robert Anderson (Mathematics, McMaster University) for pointing out a flaw in an earlier (unpublished) version of this definition.

[28] Patrick Suppes, 'The Structure of Theories and the Analysis of Data', pp. 266–83, in F. Suppe (ed.), *The Structure of Scientific Theories* (Urbana: University of Illinois Press, 1974); the quotation is from p. 277.

[29] Because the relevant literature is extensive I shall only list some of my own writings and some along related lines of approach, and refer to the references therein. See my 'Meaning Relations Among Predicates', *Noûs*, 1 (1967), 160–79; 'Meaning Relations, Possible Objects, and Possible Worlds' (with K. Lambert), pp. 1–20, in K. Lambert (ed.), *Philosophical Problems in Logic* (Dordrecht: Reidel, 1970); 'On the Extension of Beth's Semantics of Physical Theories', *Philosophy of Science*, 37 (1970), 325–39, and 'A Formal Approach to the Philosophy of Science', pp. 303–66 in R. Colodny (ed.), *Paradigms and Paradoxes: The Philosophical Challenge of the Quantum Domain* (Pittsburgh: University of Pittsburgh Press, 1972); R. Stalnaker, 'Anti-Essentialism', *Midwest Studies in Philosophy*, 4 (1979), 343–55. Gary Hardegree, 'Reichenbach and the Logic of Quantum Mechanics', *Synthese*, 35 (1977), 3–40; Linda Wessels, 'Laws and Meaning Postulates', in R. S. Cohen *et. al.* (eds.), *PSA 74* (Boston: Reidel, 1976), and Ronald Giere, *Understanding Scientific Reasoning* (New York): Holt, Rinehart, Winston, 1979), Ch. 5, 'Theories'.

[30] The main ones are by Robert Stalnaker, Richmond Thomason, and David Kaplan, mostly unpublished; see R. Stalnaker, 'Assertion', *Syntax and Semantics*, 9 (1977), and 'Pragmatics', in G. Harman and D. Davidson, *Semantics of Natural*

Language (Dordrecht: Reidel, 1972) pp. 380–97 (see also n. 36 to Ch. 5 above).

[31] There is of course much more to the semantic analysis of modality and its relation to metaphysics than I have indicated here. See also my 'The Only Necessity is Verbal Necessity', *Journal of Philosophy*, 74 (1977), 71–85, and 'Essence and Existence', pp. 1–25 in N. Rescher (ed.), *Studies in Ontology, American Philosophical Quarterly Monograph No. 12* (Oxford: Blackwell, 1978). There have recently been many flirtations with modal realisms and neo-Aristotelian essentialism in philosophy of science. For references and a critique see Hugh Mellor, 'Natural Kinds', *British Journal for Philosophy of Science*, 28 (1977), 299–312, and Sir Alfred Ayer, 'Essentialism' presented at the Symposium on Levels of Reality, Florence, September 1978, to be published in the proceedings.

CHAPTER 7

[1] Except for minor changes, this chapter is the same as 'Theoretical Entities: The Five Ways', *Philosophia*, 4 (1974), 95–109. My reason for writing it was the remark of an eighteenth-century wit that everyone believed in the existence of God until the Boyle lectures proved it.

[2] The exposition of scientific realism on which I draw mainly is that in Ch. II of J. J. C. Smart, *Philosophy and Scientific Realism* (London: Routledge and Kegan Paul, 1963). The number of such expositions is currently growing in a geometric progression, however, and it may not be too optimistic to hope that scientific realism will soon be a widely accepted philosophical dogma.

[3] W. Paley, *Natural Theology* (1802); abridged edn. by F. Ferré (Indianapolis: Bobbs–Merrill, 1963). See also n. 15 below.

[4] C. G. Hempel and P. Oppenheim, 'Studies in the Logic of Explanation', *Philosophy of Science*, 15 (1948).

[5] J. J. C. Smart, op. cit., p. 32.

[6] A. Kenny, *The Five Ways* (London: Routledge and Kegan Paul, 1969), pp. 43f.

[7] I am adapting an argument introduced in another context by R. H. Thomason and wish to thank R. de Sousa for pointing out an earlier technical error.

[8] J. H. Hick, *Arguments for the Existence of God* (London: Macmillan and Co., 1970), p. 44.

[9] C. S. Peirce, *Essays in the Philosophy of Science* (Indianapolis: Bobbs–Merrill, 1957; ed. V. Thomas), p. 166; from his 'The Reality of Thirdness'.

[10] J. J. C. Smart, op. cit., p. 39; words in square brackets are added.

[11] The reference is to Aristotle, *Metaphysics* II, 1 (993^{b}30).

[12] This line of argument was suggested to me, though not for this purpose, by Nancy Cartwright.

[13] J. J. C. Smart, op. cit., p. 47.

[14] Ibid., p. 49.

[15] J. H. Hick, op. cit., pp. 2–7.

[16] Ibid., p. 48.

[17] The article initiating these arguments, now deservedly a classic in philosophy of science, is by G. Maxwell, 'The Ontological Status of Theoretical Entities', in H. Feigl and G. Maxwell (eds.), *Minnesota Studies in the Philosophy of Science*, Vol. III (Minneapolis: University of Minnesota Press, 1962). I wish to thank Professors Maxwell and Smart and many others for their indulgent response to this paper.

[18] I owe the example to Margot Livesey.

Index